TERRORISM AND DISASTER: NEW THREATS, NEW IDEAS

RESEARCH IN SOCIAL PROBLEMS AND PUBLIC POLICY

Series Editors: William R. Freudenburg and
Ted I. K. Youn

Recently Published Volumes:

Volume 1: 1979

Volume 2: 1982

Volume 3: 1984

Volume 4: 1987

Volume 5: 1994

Volume 6: 1997

Volume 7: 1999

Volume 8: The Organizational Response to Social Theory – edited by Stephanie W. Hartwell and Russell K. Schutt, 2000

Volume 9: Environmental Risks: Perception, Evaluation and Management – edited by G. Böhm, J. Nerb, T. McDaniels and H. Spada, 2001

Volume 10: The Environmental State Under Pressure – edited by Arthur P. J. Mol and Frederick H. Buttel, 2002

RESEARCH IN SOCIAL PROBLEMS AND PUBLIC POLICY
VOLUME 11

TERRORISM AND DISASTER: NEW THREATS, NEW IDEAS

EDITED BY

LEE CLARKE
Department of Sociology, Rutgers University, New Jersey, USA

United Kingdom – North America – Japan
India – Malaysia – China

Emerald Group Publishing Limited
Howard House, Wagon Lane, Bingley BD16 1WA, UK

Reprints and permission service
Contact: booksandseries@emeraldinsight.com

British Library Cataloguing in Publication Data
A catalogue record for this book is available from the British Library

ISBN: 978-0-76-231043-2
ISSN: 0196-1152 (Series)

CONTENTS

LIST OF CONTRIBUTORS *vii*

INTRODUCTION: 9.11 AS DISASTER: ON WORST CASES, TERRORISM, AND CATASTROPHE
Lee Clarke *1*

A CIVIL DEFENSE AGAINST TERROR
Orlando Rodriguez *7*

EMPIRE OF FEAR: IMAGINED COMMUNITY AND THE SEPTEMBER 11 ATTACKS
Ann Larabee *19*

DISASTER BELIEFS AND INSTITUTIONAL INTERESTS: RECYCLING DISASTER MYTHS IN THE AFTERMATH OF 9–11
Kathleen Tierney *33*

THE FOX AND THE HEDGEHOG: MYOPIA ABOUT HOMELAND SECURITY IN U.S. POLICIES ON TERRORISM
James K. Mitchell *53*

TERRORISM AS DISASTER: SELECTED COMMONALITIES AND LONG-TERM RECOVERY FOR 9/11 SURVIVORS
Brent K. Marshall, J. Steven Picou and Duane A. Gill *73*

RECONSIDERING CONVERGENCE AND CONVERGER LEGITIMACY IN RESPONSE TO THE WORLD TRADE CENTER DISASTER
James M. Kendra and Tricia Wachtendorf 97

CONCEPTUALIZING RESPONSES TO EXTREME EVENTS: THE PROBLEM OF PANIC AND FAILING GRACEFULLY
Lee Clarke 123

LIST OF CONTRIBUTORS

Lee Clarke	Department of Sociology, Rutgers University, New Jersey, USA
Duane A. Gill	Social Science Research Center, Mississippi State University, Mississippi State, USA
James M. Kendra	Department of Public Administration, University of North Texas, Texas, USA
Ann Larabee	Department of American Thought and Language, Michigan State University, East Lansing, USA
Brent K. Marshall	Department of Sociology and Anthropology, University of Central Florida, Orlando, USA
James K. Mitchell	Department of Geography, Rutgers University, New Jersey, USA
J. Steven Picou	Department of Sociology and Anthropology, University of South Alabama, Alabama, USA
Orlando Rodriguez	Department of Sociology/Anthropology, Fordham University, New York, USA
Kathleen Tierney	Department of Sociology and Natural Hazards Center, University of Colorado at Boulder, Boulder, USA
Tricia Wachtendorf	Disaster Research Center, University of Delaware, Newark, USA

INTRODUCTION: 9.11 AS DISASTER: ON WORST CASES, TERRORISM, AND CATASTROPHE

Lee Clarke

Things will never be the same, some say, because of 9.11. We feel more vulnerable, more threatened, more at risk. It was the worst terrorist attack on U.S. soil, goes the refrain. It was dramatic beyond our worst nightmares. Like millions of others, I watched the events of that lovely morning unfold on television. When the South Tower fell for a few seconds I could not see it collapsing. My blindness wasn't because of the smoke and dust. It was a cognitive blindness. I could not believe my eyes and so, somehow, my mind denied my brain the truth of the moment.

The 9.11 attack – especially the collapse of the World Trade Center – *was* a worst case. One of the attributes of a worst case is that it's overwhelming. That's why we say things will never be the same. The attacks were overwhelming in a number of ways. They overwhelmed our imaginations as we watched airliners crashing. It was instant death on a scale that few people outside of war have ever witnessed. The attacks overwhelmed our political and organizational systems too. In the wake of 9.11, and the anthrax poisonings later that Fall, it became painfully evident how ill-prepared our organizations were to either prevent or respond to attack and disaster. The list of failures is long, embarrassing, and massive. America's intelligence and security organizations had plenty of information and warning but were poorly organized and in some cases patently incompetent. None of the 19 hijackers were even detained at the airports, even though some of them were carrying knives and didn't have proper paperwork. While it's wrong

Terrorism and Disaster: New Threats, New Ideas
Research in Social Problems and Public Policy, Volume 11, 1–6

ISSN: 0196-1152/doi:10.1016/S0196-1152(03)11001-0

to say that all the horror could have been prevented, clearly some of it should have been.

As bad as it was, it could have been worse. That, too, is an attribute of worst cases – they can always be worse. The planes could have been full of people. The planes could have crashed into the WTC an hour later, when the buildings would have had many more people in them. The people in the plane that crashed in Pennsylvania could have failed and that plane could have made its way to the White House, or the Capitol building.

It has become cliché to ask what we have learned from 9.11. The reason that it's cliché is that asking the question comes to sound rather like asking children what they've learned from their latest mistake. Just as children give parents answers they think the parents want to hear, so the "lessons learned from 9.11" come to sound homiletic, too easy, even vacuous. Nonetheless, here are some things that we *should have* learned by now:

- People are resilient in the face of catastrophe;
- People hardly ever panic in disasters;
- Americans are despised in many parts of the world;
- We won't be safe from terrorism by chest-thumping and bombing people;
- Our airports are vulnerable;
- Our critical infrastructure is vulnerable.

All of those things were known well before 9.11 by academics who know about terrorism and disasters. That's another reason that talk about "lessons learned" is often so much hot air. The lessons are *already there* but elites have to pay attention if they are to matter. But even after over 3,000 deaths, and the property destruction, elites seemed not to have learned very much from the attacks. They still think people are prone to panic, even though disaster researchers have known for a long time that is false. As a result, political leaders and even professional disaster planners, who ought to know better, create response plans that are based on faulty assumptions. Worse, the chief response to all the intelligence failures has been to create a monster bureaucracy with unprecedented powers to infringe on people's civil rights. (The coincidence of the attacks with the advent of a far-right federal government lead to major compromises in people's privacy rights.) But the terrorists are organized in small networks so it's unlikely that creating a big organization will do much more than merely assure the public that *something* is being done about the problem.

All worst cases have silver linings, and so does 9.11. At the time of the attacks Mayor Rudolph Giuliani was facing a hostile public and press over his high-handed dealings and his disastrous personal life. After 9.11 Giuliani was a hero, and will now be remembered that way. At the time of the attacks President George Bush was still reeling from having the Supreme Court hand him an election in which he lost the popular vote by a significant margin. He was also being doubted because of his lack of vision regarding public policy. After 9.11 President Bush was seen as the defender of the free world, having crushed Afghanistan and the Taliban in fairly short order, only to set up to overthrow Iraq.

Another silver lining of 9.11 is that it gave academics, including yours truly, something new to write about. This volume is one such product. In November 2001 Barbara Reskin, the president of the American Sociological Association, and Felice Levine, the ASA executive officer, made it possible for me to organize a special session at the 2002 Chicago meetings on "The 9.11 Terrorist Attacks as Disasters." I then recruited a number of people to speak at the session. The final roster included Tom Kneir, Special Agent in Charge at the FBI, in Chicago, Andrew Greeley from Universities of Chicago and Arizona, Kathleen Tierney from the Disaster Research Center at the University of Delaware, William Freudenburg from Universities of Wisconsin and California-Santa Barbara, and myself as moderator.

The session was so well attended that people were sitting on the floor. The speakers were dynamic and interesting, although the FBI representative didn't say anything surprising. But the others certainly did. Tierney drove home two points: that panic is rare and that command and control models are the wrong ones to follow if you're preparing for disaster. Greeley looked at data on religiosity and religious commitment following 9.11 and found *nothing* – people didn't go to church more, nor did churches get new members. The events of 9.11 had no significance for religion in the United States. Freudenburg talked about some political and organizational aspects of disaster response. The session couldn't have been more relevant, which is not always true of sessions at the ASA meetings.

Some time later, Bill Freudenburg asked me to edit this special issue of *Research on Social Problems and Public Policy*. I asked all those who participated in the ASA session to contribute. Kneir and Freudenburg declined. Greeley thought his non-findings weren't all that interesting and I failed to convince him otherwise. In the meantime I asked a number of other people to submit something and the results appear in this volume. It was a terribly demanding production schedule, because we wanted to get the volume out as quickly as possible. Our authors are to be commended for producing high quality work under quite pressing time constraints.

You will notice that all the work here is not of a type. I solicited work from people in the humanities as well as in social science. If we are to gain any illumination from 9.11 it must come from several sources. Certainly, the social sciences are not the only legitimate source of knowledge about 9.11. That Freudenburg and Ted Youn, who usually co-edit this journal, encouraged this unorthodox approach attests to their intellectual flexibility.

One of the things we all want to know is whether we're safer now than on September 10, 2001. As a group, our authors are pessimistic that we are. They don't think that actions now being taken by those who claim to protect us will do much good, and they do think that survivors will continue to suffer far into the future. Our authors also do not shy away from what we *should* be doing. That's all to the good. It is foolish to withhold our advice, as long as we're clear about the limits of what we know.

One issue no one raises is that there have been no attacks in the U.S. since September 11, 2001. Many of us thought there would have been more, certainly within the first year. The issue is hard to address because of a paucity of evidence, and because of national security concerns. Clearly one possibility is that Al Qaeda networks have been sufficiently disrupted that they've not been able to mount another attack. If that were the case then we can only hope that our leaders are learning the lessons from the disruptions. There is no reason to be confident that is the case.

This volume, like all edited collections, is not perfectly coherent. I don't see how it could be so. My goal was not to present a coherent work but to gather together in one place some smart people with interesting things to say about the events of 9.11 or about larger questions, concerning disaster, posed by those events. I think the effort has been successful, and I hope you do too.

Orlando Rodriguez and Phyllis Schafer Rodriguez's son died in one of the collapsing towers. His name was Greg. Two months after 9.11, Orlando was asked to give a speech at Fordham, where he is a professor, on the meaning of civil defense and the proper response to the attacks. His lecture, in the main, took the form of asking important questions. He's expanded that lecture considerably for this volume. His intent is to make us think through our values and what those values entail for responding to the terrorists. The Rodriguez' implore political elites not to commit further violence in their son's name. I'm not sure I could do the same.

Ann Larabee tells us a cultural story about the stories that are told about 9.11. Fear is now part of the social fabric, she says, in new ways. Our institutions and leaders were revealed by the attacks to be inadequate to the task of protecting us. Their legitimacy was threatened, although it's hard to find evidence of that in opinion polling regarding institutional trust and confidence. In response, leaders now assert more control over the world than they actually command, setting themselves up for embarrassment later on when their promises will be, again, shown to be hollow. The new empire of fear colors nearly everything we see, and Larabee points out that pervasive fear fits well with the urge to command and control. Crisis management now becomes an excuse to stifle dissent and narrow arguments about public policy. In that way the worldview of the terrorists is quite similar to the worldview of key policymakers in the Bush administration.

Kathleen Tierney has spent a great deal of time since 9.11 debunking disaster myths. She is uniquely qualified to do it, having served as the Director of the Disaster Research Center, at Delaware, and now as Director of the Natural Hazards Center, at the University of Colorado-Boulder. One reason such debunking is important is that disaster myths are not politically neutral, but rather work systematically to the advantage of elites. Elites cling to the panic myth because to acknowledge the truth of the situation would lead to very different policy prescriptions than the ones presently in vogue. The chief prescription is, she notes, that the best way to prepare for disasters is by following the command and control model, the embodiment of which is the federal Department of Homeland Security. Thus do panic myths reinforce particular institutional interests. But it is not bureaucrats who will be the first-responders when the next disaster, whether brought by terrorists or some other agent, comes. It won't even be the police or firefighters. It will be our neighbors, it will be the strangers in the next car, it will be our family members. The effectiveness of disaster response is thus diminished to the degree that we over-rely on command and control. This is another case where political ideology trumps good scientific knowledge about how the world works.

Ken Mitchell talks of the "hard won lessons" that are being ignored in the political and organizational response to 9.11. He points out that defense is trumping real security, and in ways that compromise not only democratic culture but also make us less safe. Certainly he's right. But we have to ask what *can* the government actually do against such things? He is clear that the things the American government is doing in the name of protecting its people are in fact putting its people in greater danger. The larger issue is that in times of great emotional turmoil there's a tendency to think about the government as a protector of us all, as per the phrase "the national interest." But it is unwise to forget, especially when emotions and patriotic fervor are running high, that there are very few single interests. Mitchell also notes that the goal of preventive efforts is directed mostly at "critical infrastructure." That means that most attention is directed at protecting things rather than people, leaving us terrifyingly vulnerable.

Brent K. Marshall, J. Steven Picou and Duane A. Gill are ambitious. They mine the literature on disasters to see what predictions might be made about the prospects of long-term recovery for 9.11 survivors. They see parallels in how people typically respond to natural disasters. But they especially draw lessons from work on technological disasters. Technological disasters, they note correctly, have been shown to be highly correlated with the development of a "corrosive" rather than an "altruistic" community. The primary reason for that is that lawsuits are more likely to be filed in technological events. That sets people against each other, which has obvious implications for possibilities of developing a sense of community and common recovery. Marshall et al. predict that the 9.11 survivors will look much like survivors of other technological disasters, which is to say recovery will be halting and possibly won't happen at all for many people.

James Kendra and Tricia Wachtendorf were scheduled to be in New York City on September 12, 2001. As researchers from the Disaster Research Center, they were going to witness a simulation of a bioterrorist attack. That simulation was upstaged by the real attack, and on September 13 Kendra and Wachtendorf went to New York, where they were welcomed into the reconstituted New York City Office of Emergency Management. They stayed in New York for two months. They thus have a unique perspective on what happened in lower Manhattan in the intermediate aftermath. One thing they observed was what they say is a new kind of convergence in disasters. Convergence – where people of all sorts are drawn toward a disaster scene – has long been noted in the disaster literature. Kendra and Wachtendorf advise disaster managers to plan for and explicitly incorporate unofficial volunteers. This is yet another way in which informal organization becomes at least as equal as formal organization in disasters. Their chief conceptual innovation is to point out the study of convergence should center on how people negotiate legitimacy in their attempts to be part of recovery.

In the Fall of 2002 I published an article in *Contexts*, the American Sociological Association's magazine directed at a general audience. That piece concerned the myth of panic in general, although I used some stories from the 9.11 attacks as evidence. The *Contexts* article was extracted and refined from a longer, more academic piece that I was working on. That article appears here. My chief point is to suggest the outlines of a theory of people's response to disaster that gets away from usual categories. Specifically, I suggest the idea of "failing gracefully" as a social structural perspective on what we usually see as panic. Behavior is embedded in social contexts of various sorts, and that central fact doesn't disappear in disaster.

This collection of articles raises more questions than it settles, but that is appropriate. We are only now beginning to fathom the deeper meanings of 9.11. It will be a long time, I think, before anyone should feel confident in their knowledge about those meanings.

A CIVIL DEFENSE AGAINST TERROR

Orlando Rodriguez

ABSTRACT

Terrorism's political strategy is to force the State to accept social or political change by inciting fear among civilian sectors of a population, while attempting to gain sympathy from other civilian sectors whose interests terrorists claim to represent. Usually, the civilian recipients of terror see themselves as defenseless without the State's intervention, and they view counter-terrorist policy as being in the province of the State. Policy analysts tacitly or explicitly share this view. In reality, civilian involvement could improve the effectiveness of counter-terrorist policy. To illustrate this, I discuss three issues around which citizen groups have organized and are making an impact on public opinion and policy in the aftermath of September 11th, 2001: public safety, particularly the Indian Point, New York, nuclear plant; learning from mistakes by government intelligence agencies in order to prevent future attacks; and rejection of war and violence as responses to terror.

INTRODUCTION

This paper originated as an attempt, after the convulsive events of September 11th, 2001, to convince a gathering of students and faculty (perhaps to convince the author more than anybody else at that meeting) that it is possible and desirable to apply reason to inherently emotional public policy questions. As such, the paper upholds the spirit of policy research. It is directed as much to policy researchers as the citizens on whose behalf policy researchers ultimately think and write.

Terrorism and Disaster: New Threats, New Ideas
Research in Social Problems and Public Policy, Volume 11, 7–17

ISSN: 0196-1152/doi:10.1016/S0196-1152(03)11002-2

In 1982, Stephen Jay Gould, a Harvard University biologist, was diagnosed with a rare and extremely deadly form of cancer. His doctor told him that the median mortality for that form of cancer was eight months. In response to his request for medical literature on his diagnosis, he was told there was nothing really worth reading. After he looked at the literature he realized what the doctor meant: the medical literature was very pessimistic about treatment and survival chances. After feeling hopeless for a short while, he then began to make use of his professional training in the service of his own survival. This helped him reach two conclusions (Gould, 1985, pp. 3–7). His first conclusion was that,

> If a little learning could ever be a dangerous thing, I had encountered a classic example. We don't know why (from my old-style materialistic perspective, I suggest that mental states feed back upon the immune system). But match people with the same cancer for age, class, health, socioeconomic status, and, in general, those with positive attitudes, with a strong will and purpose for living, with commitment to struggle, with an active response to aiding their own treatment and not just passive acceptance of anything doctors say, tend to live longer.

His second conclusion was that,

> When I learned about the eight-month median, my first intellectual reaction was: fine, half the people will live longer; now what are my chances in being in that half. I read for a furious and nervous hour and concluded with relief: dammed good. I possessed every one of the characteristics conferring a probability of longer life . . . I knew how to read the data properly and not despair . . . I immediately recognized that the distribution of variation about the eight-month median would almost surely be what statisticians call 'right skewed' . . . The distribution of variation had to be right skewed, I reasoned. After all, the left of the distribution contains an irrevocable lower boundary of zero (since mesothelioma can only be identified at death or before). Thus, there isn't much room for the distribution's lower (or left) half – it must be scrunched up between zero and eight months. But the upper (or right) half can extend out for years and years, even if nobody ultimately survives. The distribution must be right skewed, and I needed to know how long the extended tail ran – for I had already concluded that my favorable profile made me a good candidate for that part of the curve.

How is the median defined and why is skewness an important characteristic of a statistical distribution? Is this important to know? If one is diagnosed with cancer, knowing this may be a life and death matter. How can citizens defend themselves against terrorism? Knowing what a median is will probably not matter, but other kinds of knowledge may be critical for our collective well-being. To effectively defend ourselves against terrorism, it is also important to take to heart Gould's first conclusion about medical knowledge, paraphrasing it slightly to adapt it to our immediate issue of concern: having a positive attitude, a strong will and purpose for living, commitment to struggle, and an active response to aiding our own survival – not just passive acceptance of anything those in authority say – will give us a better chance of successfully dealing with terrorism.

Keeping these two principles in mind, how can citizens defend themselves against the terrorist threats we face today? There are three complementary approaches:

- We can protect ourselves by taking those measures that it is in our power as individuals to take. In times of active threat, we can avoid areas likely to be targets of future attacks, such as theaters, mass transit, malls. We can also store supplies of food, drink and medicines in case a future disaster strikes us;
- We recognize that there are obvious limitations to our protecting ourselves as individuals, so we look to government for protection. Since we live in a democratic society, this is done on our behalf and with our approval, tacit or explicit;
- The third approach to protecting ourselves from future terror is by making use of the political institutions of civil society.

To understand why civil society is needed, and is a critical component of the three defenses against terrorism, it is important to reflect on the meaning of terror. Consider whether these ideas are useful in helping us to evaluate the actual threat that terror represents:

From Merriam Webster's *Collegiate Dictionary* (1998), 10th edition:

> **terror**, *n* [*terreur*, from Latin *terror*, to frighten; akin to Greek *trein* to be afraid, flee; *tremein* to tremble]. A state of intense fear; a frightening aspect; a cause of anxiety; worry; violence (as bombing) committed by groups in order to intimidate a population or government into granting their demands. **terrorize**, *vt*. To fill with terror or anxiety; to coerce by threat or violence.

We fear terror because we can visualize the deaths of our loved ones and ourselves in future acts. Our fear is fed by the realization that terrorist acts are difficult to prevent. Terrorists can evade surveillance, and keep us off-guard through changes in tactics and targets. Technological developments have increased weapons' deadliness. The armaments industries of advanced countries have made access to weapons easier now than ever before. These developments have made citizens in many countries at greater mortal risk today than ever before.[1] The protracted conflicts of Srilanka, Northern Ireland, Israel, Palestine, and Spain, to name a few countries, make us realize that it may be more realistic to think in terms of reducing terrorism rather than eliminating it. If terrorism is eliminated, deaths are prevented. If terrorism is reduced, deaths are minimized but not avoided. So a realistic counter-terrorism policy brings with it a realistic fear of death.

People die from terrorist acts, and the fear is real. But terrorism is theater, and its drama makes us inordinately susceptible to it. We have been trained to understand through dramatic images. Graphic TV images are more effective than un-dramatic print. We need to recognize that the symbolic aspect of images is an important aspect of terror. It is very important to understand that while the threat to life is real, fear makes us misunderstand and overestimate the actual threats to life that terrorism represents. It is also important to understand that the media can facilitate terrorist aims. However, the media can also facilitate the State's counter-terrorist aims, since news can and has been used by governments for propaganda purposes.

Since the military and the criminal justice system are often the lead sectors chosen by the State to combat terrorism, terrorism can bring forth a violent response that can threaten our lives and other populations' lives, as well as our civil liberties. Terrorists view the loss of human life as regrettable but serving a higher purpose, an ideal that is more important than mere human life. Counter-terrorist policy that stresses violent response often makes us fall into the same mind-set. Did some innocent people die in a military counter-terrorist operation in some targeted country? "Regrettable," our leaders would say, "but terrorism must be wiped out." We heard talk about "collateral damage" in dispatches from the Afghan military operations. This mentality only feeds upon terrorists' political aims because the killing of innocent bystanders to the conflict can be used to reinforce their justification for violence.

The stronger the civil libertarian rules and values that are part of the political culture, the more measured the response to terrorism. Compare the loss of life and curtailment of human rights in European countries facing left-wing and right-wing terror in the 1970s, with the same damage in countries like Argentina and Uruguay, where during the same decade the military killed far more people than the terrorists (Adams, 2000, pp. 68–69). This is not to say that countries like ours are immune from threats to civil liberties. For example, two months after September 11th, the *New York Times* (Rutenberg & Carter, 2001, B2) reported,

> As television news networks cover the war, they are increasingly coming under criticism from conservatives who say they exhibit a lack of patriotism or are overly negative toward the government. Although the networks say the criticism is not affecting their coverage, they are having to sort through basic issues like whether patriotism can coexist with objectivity and how to raise tough questions about the policies of an administration that is receiving overwhelming popular support. Much of the criticism comes from a group of conservative media voice and outlets, including Rush Limbaugh's radio talk show, The New York Post's editorial page, The Drudge Report, and some commentators on the Fox News Channel. Much of the information for their critiques has been assembled by a conservative media watchdog organization called the Media Research Center, which hires full-time monitors to watch the network newscasts . . . One executive said that while there had always been conservative critics, the current criticism could have a more volatile effect because of the fears and sentiments of the audience. Twice in recent days, networks made decisions, at least in part, to smother accusations that they lacked patriotism or were skewing coverage toward the enemy. CNN decided to require that reports of civilian casualties in Afghanistan be balanced with reminders of the Sept. 11 toll.

Thus, government pressure on the media to produce anti-terrorist propaganda, combined with threats to civil liberties, led Perl (1997, cited in Adams, 2000, p. 145) to conclude,

> The potential power of these [terrorist] groups seems to lie not in their threat to overthrow society by force of arms per se, but in their ability to symbolize the fragility and vulnerability of the social order and force that order to subvert itself by eroding the liberal and democratic values upon which its own legitimacy is based.

The recipients of terror – the victims and survivors – are not the only audience, perhaps not even the principal audience for terror. The principal audience may be those who nurse grievances against our policies, those who the terrorists claim to represent. Terrorist tactics do not work when the people terrorists supposedly represent turn against violence. By the same logic, terrorist attacks like the one we suffered on September 11th somehow speak to the grievances of people in other countries. We know that many disenfranchised people in poor countries have a moral vision that sees killing for what it is. But many feel vindicated by the deaths of the "other." Difficult as it is for terror's survivors to accept, and in the interest of our own survival as a free people in a mutually dependent world, we must learn to understand why many among the disenfranchised cheer terrorists on. This entails devoting our energies to understanding our foreign policy and the effects of our global economy on people all over the world. Citizens are not used to paying attention to these issues, so we often leave them to experts. But we must work harder to understand the world now, as we did during the time of the Vietnam War.

An effective defense involves civil action in the multiple meanings of the word "civil." Again from Merriam Webster's *Collegiate Dictionary*:

> **civil**, *adj* [from Latin *civilis*]. Of or relating to the state or its citizenry; involving the general public, their activities, needs, or ways, or civic affairs, as distinguished from special (as military or religious affairs); adequate in courtesy and politeness.
>
> **civil defense**, *n* The system of protective measures and emergency relief activities conducted by civilians in case of hostile attack, sabotage, or natural disaster.
>
> **civil liberty**, *n* Freedom from arbitrary governmental interference (as with the right of free speech) specifically by denial of governmental power and in the U.S. especially as guaranteed by the Bill of Rights.

Civil means relating to citizens. The counterpart of citizens can be seen as military or government officials. Civil also means "involving the general public, their activities, needs, or ways or civic affairs, as distinguished from special (as military or religious affairs)." An important implication of this is that we should strengthen and promote the role of civil society in a campaign against terror. Part of this task is to clarify the legitimate and illegitimate roles of government in curtailing terrorism, its appropriate and inappropriate functions, its limitations and advantages. The government and its critics should also be civil as in "adequate in courtesy and politeness," that is, being civil to opponents and thus helping to protect our civil liberties.

We might conclude from the preceding arguments that an effective civil defense against terror involves taking an active role in questioning our government's response to the threat of terror. There has been citizen action around three concerns that illustrate types of questions that concerned citizens might have asked themselves as they tried to come to grips with their feelings about government anti-terror policy. I have chosen three examples of issues around which citizen groups have successfully organized and are making an impact on public opinion and policy in the aftermath of September 11th, 2001: public safety, particularly the Indian Point nuclear plant; learning from mistakes by government intelligence agencies in order to prevent future attacks; and rejection of war and violence as responses to terror.

Organizing for Public Safety

How prepared is our government to deal with other terrorist threats? We know about what has been done about airport security and we know there is disagreement in high circles about whether the government should spend the resources needed to protect air travel. How reasonable is our government's approach? Are there other threats that we should worry about? Should we worry about threats to our bridges, to our nuclear energy plants, to shopping facilities and mass transit? Should we worry about chemical threats such as the Sarin nerve gas attack on the Japanese public a couple of years ago? If so, what should the government now do about this?

The attacks on the World Trade Center killed approximately 3,000 people. Very soon after that, many New Yorkers realized that had the attackers chosen to crash two airplanes into the Indian Point nuclear energy plant, located 35 miles north of New York City in Northern Westchester County, the death toll created by a nuclear meltdown could have well been in the hundreds of thousands, not to mention the paralyzing effect a nuclear accident would have had on the metropolitan area's economy. In response to these fears, New York citizen groups, such as New York Public Interest Research Group (NYPIRG), Riverkeeper and Citizens Concerned about Indian Point, have aimed to shut down the Indian Point plants in the long run, and in the short run, press facility operators and the Nuclear Regulatory Commission (NRC) to develop more effective safety and security regulations. A coalition of public interest, environmental and grass roots groups has been quite vocal in the past year. Its members include NYPIRG, Riverkeeper, Hudson River Sloop Clearwater and the Westchester Citizens Awareness Network. Their website is informative and action oriented. A Marist Poll on Workability of Regional Evacuation for Indian Point Nuclear Plant released by Riverkeeper, found that many citizens and government officials "consider the plant a prime potential terrorist target in the wake of the September 11 World Trade Center attack" and lack confidence in the evacuation plan. "What Can You Do Now?" lists ten things individuals can do to shut the plant, such as contacting New York State Gov. Pataki to protect public health and safety by shutting it down, writing to the Regional Director for the Federal Emergency Management Agency (FEMA) urging him to rule that the plan is unworkable, familiarizing oneself with the issue, and telling friends and neighbors to join a member organization of the coalition. By now, television viewers and suburban commuter railroad passengers are accustomed to seeing ads calling for the shutting down of Indian Point. As a result of citizen action, NRC officials and state and local politicians have had to devote time and energy to counter citizens' demands, including Gov. Pataki's reelection campaign. Today, Indian Point is still vulnerable to a terrorist attack, especially from the air, but New York citizens' safety from such attacks has marginally improved. Most important, citizens involved in the campaign against Indian Point feel a sense of entitlement and empowerment much different from the initial post-September 11th fears about their families' safety. They are not satisfied with merely focusing on personal solutions, such as stocking water and food supplies in car trunks and keeping potassium iodide pills on hand (Rowe, 2001).

Learning from the Intelligence Failure to Prevent Future Terror

How forthright have our government intelligence agencies been in learning from the September 11th attack to reorganize their intelligence gathering? Will current efforts, like the creation of a cabinet-level Homeland Security Department, really lead to more effective intelligence gathering and therefore greater public safety? Or are these really inter-agency agreements to deflect blame and protect existing turf? What can citizens do to influence obtaining real and public knowledge of the September 11th intelligence failure? No aspect of the government's post September 11th policy has been as controversial as its explanation, or lack of an explanation, for how warning signs and intelligence sources were ignored by the Pentagon, the Federal Bureau of Investigation (FBI), the Central Intelligence Agency (CIA), the National Security Agency (NSA), among others. As shown in a recent New York Times article (Hulse, 2002):

> The idea of an independent commission like those that investigated the attack on Pearl Harbor and the assassination of President John F. Kennedy was broached soon after the attacks. The White House initially objected, saying efforts devoted to the inquiry could detract from the

> fight against terror. The administration did ultimately agree to a joint inquiry by the House and Senate intelligence committees, viewing that approach, Congressional Republicans say, as more manageable because much of it would be conducted in secret and Republican lawmakers would share in lead roles.
>
> But the leaders of the joint committee conducted an aggressive inquiry that uncovered previously undisclosed warnings of the attacks, and those results gave momentum to the push for a broader investigation and to the White House's dropping its objections.
>
> By late September, the Senate had followed the House in voting to approve an independent commission. But their versions differed, and negotiations among the two chambers and the White House became tortured. Many Democrats say that the White House strategy was to slow the very investigation that it had endorsed.

Much of the credit for the White House ultimately acceding to an independent commission is due to pressure groups created by survivors and relatives of victims of the terror attacks, such as September 11th Advocates. As the *New York Times* story cited above indicated, September 11th Advocates was very influential in pressuring the White House and Congress to stop stalling the formation of a commission and is part of Families of September 11th, Inc., a coalition of victims' relatives groups, including firefighters, police, airline employees, etc., Voices of September 11th, and the Skyscraper Safety Campaign. Areas of commonality have been stressed by the coalition's organizers in order to be maximally effective, although there are undoubtedly differences of opinion and areas of concern within such a large, diverse group. Each is concerned with specific issues, such as how Ground Zero will be memorialized, the scope of compensation for relatives, opposition to the bombing of Afghanistan – and now to unilateral military action against Iraq. A campaign to require television networks to include warnings regarding violent content and images when aired was also initiated by members of the coalition. Much like the families of the Lockerbie bombing victims, who were ultimately effective in pressing the U.S., Lybia and other governments to prosecute the Lybian intelligence agents who organized the bombing (and coincidentally contributed to governments accepting the principle of international criminal trials for terrorists), September 11th relatives' groups will continue to press for an independent inquiry that generates effective and public knowledge about how to organize ourselves against terror.

Much of the negotiation content has concerned make-up and control of the commission. The White House insisted, and got its way, in naming the commission's chairperson, while Congress would then choose other commission members. After the commission was set up, there were immediate objections to naming ex-Secretary of State and ex-head of the National Security Council Henry Kissinger as Chairperson. Today September 11th relatives' groups continue to be involved in this inquiry, having expressed disappointment at the White House's naming Kissinger (Firestone, 2002), also expressing criticism of the subsequent appointment of former New Jersey Governor Thomas Kean as Chairperson after Kissinger resigned.

Rejecting War as a Way to End Terror

A final set of questions, perhaps the most difficult for us to confront as citizens, concerns our government's military attack on the ruling government of Afghanistan in reaction to the terrorist attacks. The aim was to force the ruling Taliban government to turn over to the U.S. government the members of the terrorist organization responsible for the attacks, on the assumption that they were being protected. Since this then is an anti-terrorist policy, it should be subject to the same evaluation as any other anti-terrorist policy. Most citizens are accustomed to the idea of pressuring the government on domestic issues, but they often view foreign policy as beyond their understanding (too complex) and having to be left to experts like policy analysts and high level advisors. Is it logical and desirable for citizens to question our government's military response to terror? If this is a logical thing to do, what alternatives to military attacks could our government have carried out to more effectively deal with the terrorist threat? For those who believe that our government's military approach is correct, what is the basis for this belief? Is it based on knowledge about the way the war is being conducted, and how our government changed its policy towards other countries – especially countries with large Muslim populations? Or is this belief based on the feeling that it is not morally or politically correct to question our foreign policy or the conduct of this military campaign? Is acquiescence in the current war approach, based on emotional feelings such as our country's honor, likely to advance our survival against terrorist threats?

As early as November 2001, Voices in the Wilderness, an organization that promotes non-violence, organized a march from Washington, D.C., to New York City, ending with a vigil in Washington Square Park. In the course of these events, several relatives of victims lost at the World Trade Center and the Pentagon met each other for the first time. Many of them (and others not on the march) had already spoken out against our government's threatening to attack Afghanistan militarily, with the inevitable deaths of innocent civilians, and had called for a non-vengeful approach to ferreting out and bringing the perpetrators to justice according to international law. Since then there have been two major demonstrations in Washington, D.C., protesting war against Iraq and many others in cities throughout the U.S. These activities did not ultimately deter the government from a war policy. But as the contending parties seek their preferred solutions, both government and citizens have the Vietnam war as a historical precedent for effective citizen efforts against war.

CONCLUSIONS

When Steven J. Gould learned that his life was endangered by a deadly form of cancer, he read all he could find on the subject and questioned the conclusions drawn by experts. He took nothing at face value, concluding that the median was not the message. He died nineteen years after being told that his death was imminent. Gould's message suggests that in order for a civil defense to work, citizens must be motivated to challenge government anti-terrorist policy. As shown by this brief account of citizen action on three fronts of the struggle against terrorism, it is difficult, but not impossible, for citizens to take an active part in anti-terrorist policy.

This paper began with an appreciation of policy research's rational approach as a means to confronting emotional public policy issues. But the arguments proffered here also imply that policy analysts should be actively involved in civil anti-terrorist policy, not just as citizens but as policy experts. Obviously, this is partly a normative issue. To help in civil defense as defined here, a policy researcher must agree with this paper's value premises. Charles E. Lindblom aptly made the case for this type of involvement when he argued (1989, p. 313):

> (The policy analyst's) version of the public interest must always – if it is to be accepted as relevant – hold closer to conventional values and conventional weights among them. It has to move toward a modal version of the public interest.
>
> Now this tendency for the public interest to be defined around conventional values and weights offers some advantages to society. It means that social researchers, because their versions are so much alike, can talk easily with each other; and in turn they can talk easily to political leaders, who in turn can talk easily with each other. They all find themselves thinking very much alike. In addition, their easy agreements may make a contribution to political stability.
>
> But another result is that dissident interests are disproportionately deprived of the services of social scientists and researchers. Dissidents stand for odd sets of values that social researchers reject as not in the public interest, but instead in the interests of no more than one or more small minorities. Is that regrettable? On the ground that most dissidents are foolish people, it is not. On the ground that all major social reforms, all great improvements in the welfare of humankind come from dissidents, to deprive them disproportionately of the services of professional social inquiry would seem to be gross mistake, perhaps better characterized by a word like *tragedy* (author's emphasis).

In the policy literature, it is common to discuss the policy researcher's role as conducting dialogue with the policy maker. In this role definition, the citizen is viewed as the person on whose behalf the policy maker acts, political issue organizations as part of the environment to be confronted, and the policy maker as the policy researcher's client. In the same essay cited above, Lindblom calls for policy researchers to address citizens' needs for information such as "the range of possible institutions and practices, of the competences and incompetences of government, and of alternative social machinery like the market system, of how to train and organize themselves for political participation, of the history of constraints on their own habits of thought" (Lindblom, 1989, p. 331). In line with Lindblom's views, this paper suggests that the policy researchers engage citizens as their clients in the interest of a civil defense against terror. For the interested and concerned policy researcher it remains to learn how to become an effective consultant for citizens.

NOTE

1. Pointing to the increases in weapons deadliness, Chalk (1999) calculates that 50,070 people were killed world-wide in terrorist attacks from 1990 to 1996, double the 28,100 who lost their lives in the years between 1970 and 1983. This is in spite of a decrease in terrorist incidents in the 1990s compared to the previous two decades (Adams, 2000).

ACKNOWLEDGMENTS

An earlier version of this paper was presented on November 13th, 2001 at a forum at Fordham University, Bronx, New York, part of a series titled "Transcending the September 11th Tragedy." I am grateful to my wife, Phyllis Schafer Rodriguez for her research assistance, editorial advice and encouragement.

REFERENCES

Adams, C. (2000). *Terrorism in the twenty-first century* (2nd ed.). Upper Saddle River, NJ: Prentice-Hall.

Chalk, P. (1999). The evolving dynamic of terrorism in the 1990s. *Australian Journal of International Affairs*, *53*, 151–167.

Firestone, D. (2002). Kissinger pulls out as chief of inquiry into 9/11 attacks. *The New York Times*, December 14, p. 1.

Gould, S. J. (1985). The median is not the message. *Discover*, June, 3–7.

Hulse, C. (2002). How a deal creating an independent commission on Sept. 11 came undone. *The New York Times*, November 2, p. A8.

Lindblom, C. E. (1989). *Democracy and market system*. Oslo: Norwegian University Press.

Merriam-Webster's Collegiate Dictionary (1998). Springfield, MA: Merriam-Webster (10th ed.).

Perl, R. (1997). *Terrorism, the media, and the 21st century*. Washington, DC: Congressional Research Service.

Rowe, P. (2001). A grass roots fight. *The New York Times*, December 6th, p. 71.

Rutenberg, J., & Carter, B. (2001). Network coverage a target of fire from conservatives. *The New York Times*, November 7th, p. B2.

EMPIRE OF FEAR: IMAGINED COMMUNITY AND THE SEPTEMBER 11 ATTACKS

Ann Larabee

ABSTRACT

The attacks of 11 September 2001 created a crisis of legitimacy for the U.S. nation state. To overcome a catastrophic event that threatened national identity, the Bush administration evoked fear as the spiritual root of patriotism and the basis of a renewed security state. The modern rhetoric of crisis management was combined with a nostalgic rhetoric of national community. In the new civil defense, all citizens were enlisted to relentlessly examine their fears so that bodies, minds, neighborhoods, and ultimately the nation state could be free of terror. These conditions led to authoritarian efforts to reach deep into citizens' private lives and purge the body politic of ill-defined invaders, damaging democratic community.

FEAR AS NATIONAL RELIGION

In the United States, after the terrorist attacks of September 11, the risk society gave way to the empire of fear. In a paroxysm of defensive nationalism, the Bush administration nostalgically returned to the militarized citadel culture of the Reagan years. Then, psychological wars were waged against an Evil Empire from under an imaginary space umbrella. But with the destruction of the World

Terrorism and Disaster: New Threats, New Ideas
Research in Social Problems and Public Policy, Volume 11, 19–31

ISSN: 0196-1152/doi:10.1016/S0196-1152(03)11003-4

Trade Center, the apocalyptic nightmares of the cold war were fulfilled at an urban Ground Zero. Daily routines, ordinary objects, and peaceful neighborhoods were menaced with hidden invaders, as terrorists filled the psychic gaps left by the old communists, who now seemed comparatively benign. The immediate resurgence of paranoiac cold war patterns shows just how embedded they still are, not only in the nation's political institutions, but also in the minds of its citizens. These patterns helped give a transcendent national shape to inchoate fears of death. Ordinarily, after a great catastrophe, institutions generate a self-protective, recuperative discourse, attempting to replace fear with hope. But to bolster its legitimacy, the Bush administration evoked fear as the spiritual root of patriotism and the basis of a renewed security state.

Benedict Anderson (1991) argues that fear and overcoming fear are essential to nationalism and nation, in the formation of what he calls "imagined community." Anderson argues that the nation, unlike other social organizations, inspires citizens to willingly die for it. The imaginary community of nation is deeply concerned with death, as witnessed in its military cemeteries and monuments. Like a religion, Anderson explains, nationalism promises a "pure" and "disinterested" collective transcendence over the uncertainties of individual suffering and death (pp. 9–12, 144). To overcome a catastrophic event that threatened national identity, the Bush administration made the willingness to die for a pure nation the negotiating terms of a post-disaster America. Citizens were to purify themselves of fear, going about their lives even at the chronic risk of death.

Ultimately, national faith, like Christian evangelical faith, promised freedom from fear of evil and damnation. In the aftermath of September 11, President Bush evoked Franklin Delano Roosevelt's 1941 State of the Union address that outlined four freedoms: freedom of speech and religion, freedom from want and fear. Despite the reference to political tradition, the emphasis was overwhelmingly on an evangelical freedom from fear through the elimination of "evildoers." As citizens struggled for the meaning of mass murder, freedom from fear rose to the status of a sacred text in a national religion. In October 2001, Rudolph Giuliani (2001), then mayor of New York, told the members of the United Nations to heed the Christian hymn, "Be Not Afraid," sung at funerals for the New York firefighters. He added, "Freedom from Fear is a basic human right. We need to reassert our right to live free from fear with greater confidence and determination than ever before ... here in New York City ... across America ... and around the world" (p. 6). The 2002 commemoration in New York began in the morning with a reading of the Gettysburg Address and the Declaration of Independence and ended with the mayor's reading of Roosevelt's four freedoms and a lighting of an eternal flame. At the anniversary of the disaster, President Bush (2002) wrote a letter to the *New York Times* in which he declared that "at a moment of great testing, the spirit of men and women in New York City, at the Pentagon and aboard Flight 93 became the spirit of our country." The trajectory from the most private grief to national honor offered the comfort and inspiration of belonging to collective, spiritual destiny, free from fear.

Necessary to a transcendent nation, the specter of evil was more than specific terrorist organizations. The threat became a shapeless terror, "an unknown threat," a "constantly evolving threat," an "often-invisible enemy, an enemy that hides in the shadows" (Bush, Security, 2002). Even though the enemy organization and its members did, in fact, have names, they were often referred to as *nameless* terror. Because mass murder threatens loss of identity, names become crucial as the victims are slowly collected from the anonymity of mass death. Just as the reclaiming of victims' identities reasserts social continuity, nameless terror represents complete social disintegration. Because it is nameless, it may come from anywhere. Thus, the evil of nameless terror has been attached to a variety of social threats. CNN terrorism expert Bruce Hoffman (1999) has argued that the news media has debased the word to mean "virtually any abhorrent act of violence that is perceived as directed against society" (p. 13). President Bush included child abusers among terrorists. Attorney General John Ashcroft (2001) expanded the definition of terrorist to "those who would destroy America," not only the suspected Al Qaeda operative, but the allegedly disloyal citizens who would "erode our national unity" (p. 707). *Defending Civilization*, a report produced by Lynne Cheney for the American Council of Trustees and Alumni (2002), condemned professors for their "moral equivocation" in questioning national conduct in the war on terrorism (p. 2). The rhetoric of spiritual vigilance against an invisible, invading enemy was thus extended to the enemy within, those steeped in moral relativism. In U.S. political history, this is a familiar pattern in which limited threats lead to a wide demonization of amorphous enemies. The quest to eliminate fear leads to a hyper-state of vigilance against the evildoers, whoever they may be.

NOSTALGIC UNITY

The spiritual vision of transparent, communal solidarity was enhanced by nostalgic references to a past when pure values existed without relativism or irony. As political and social institutions look to the past for legitimacy, they erase it in a wave of nostalgia, for no such time ever existed. Such amnesias, as Anderson (1991) notes, always accompany shifts in national consciousness (p. 204). In its deployment of Roosevelt's four freedoms, the Bush administration placed itself in an idealized history as well as a sacred moment. Nostalgic patriotism resonated across a national discourse that did not brook dissent. After

September 11, with support from the government, the Ad Council (2002) sponsored several campaigns, including I Am an American and Live Brave, to "help Americans remain strong throughout the war on terrorism, just as it had during World War II." The Ad Council was formed in 1942 as the War Advertising Council, responsible for "Loose Lips Sink Ships." With Campaign for Freedom, the Ad Council created imaginary scenarios in which freedoms are taken away, very reminiscent of the Cold War's civic education dramas in which the Soviets march into small American towns and tear down the flags and close the churches. Similarly, one of Campaign for Freedom's ads begins with a seemingly ordinary church service, but the worshippers then fearfully emerge from what turns out to be a basement, the image visually quoting fallout shelters. Another ad features police officers arresting a young man in a library after he asks the stern librarian for a list of forbidden books. A third follows a shopping cart wheeling down an aisle, followed by the caption Freedom + Competition + Possibilities. All the ads are followed by the question, "What If America Wasn't America?" When Campaign for Freedom was launched, the industry publication *Advertising Age* explained that the creators, including Phil Dusenberry who once worked for Ronald Reagan's political campaign, intended to show that "people in other places want to take Americans' freedom away" (Teinowitz, 2002, p. 3). Who these people are remains a mystery, though their agents seem to be U.S. librarians and police officers. Ellis Verdi, president of the advertising agency responsible for some of the spots, explained, "You almost have to go beyond just getting a feeling. You almost have to try to create a chill. You're trying to get under people's skin a lot deeper" (National Public Radio, 2002). In theory, fears from the twilight zone will channel into national pride and appreciation of freedom.

Although the Ad Council perceived its ads as uncontroversial civics lessons, they intersected with passage of the Uniting and Strengthening America by Providing Appropriate Tools Required to Intercept and Obstruct Terrorism Act, popularly known as the USA. Patriot Act. The USA. Patriot Act, along with amendments to the Attorney General's Guidelines on General Crimes, Racketeering Enterprise, and Terrorism Enterprise Investigations, make it possible for federal agents to acquire the reading records of library patrons and spy on religious groups. These investigations can be conducted based on speculation rather than any evidence of criminal conduct. Made patriotic by the war on terrorism, fear and suspicion were given greater latitude in spurring investigations, including secret collection of information. Although the Ad Council spots reinforced the generalized fear of mysterious enemies, it also inadvertently suggested that the enemy within might be the FBI. But most U.S. citizens supported extreme measures to eliminate their fear.

A widespread, nostalgic longing for security was evident in the renewed popularity of Norman Rockwell's WWII liberty bond poster, *Freedom from Fear*, one of four based on Roosevelt's speech. After September 11, the poster became part of a broad institutional and media campaign to provide Americans with inspirational, patriotic, and psychically healing messages. It was visually quoted in advertisements and artworks, discussed in newspaper columns and commemorative speeches, and featured in the Smithsonian Museum of American History. It drew interest as part of a major retrospective of Rockwell's work, which traveled across the country, including a stint at the Guggenheim Museum in New York. And the *New York Times* used the image in a promotional campaign, changing the newspaper in the image from the *Bennington Banner* (Vermont) to the *Times*, and, in case anyone missed the connection, altered its headline to refer to the September 11 attacks (Cooper, 2002). Norman Rockwell is well known as one of the most nostalgic and sentimental of American artists, presenting an uncomplicated view of a normative national identity. But *Freedom from Fear* is a troubling image.

The poster shows two parents standing protectively over the bed of their sleeping children, the mother bending down to tuck the blanket. The father stands nearby holding a newspaper in his hand, the few visible words of the headlines announcing, "bombings kill" and "horror hits." A black, slanted shadow strikes down in the backdrop, threatening the white, nuclear family. A rag doll sprawls like a dead child on the floor. The emotional intensity of the image vacillates between the central reassurance of parental care and the claustrophobic periphery of the dark room with its amorphous threats. In the tension between love and fear, Harvard law professor Richard Parker (2002) finds "the energy, the yearning, that is the inward experience of American patriotism" (pp. 422–423). If so, than that inward experience is given shape by simple dichotomies between protectors and protected. American citizens are asked to be both like the parents, focusing their attentions inward to the nuclear family, and like the sleeping children, giving themselves over to caring, knowing guardians. Protection operates in a tightly enclosed world, in which the only alternative to fear is vigilant, compassionately conservative authority figures.

Among the troubling aspects of freedom from fear was that it legitimated a strongly authoritarian impulse in government, which attempted to reach deep into its citizens' psyches. From homes to bodies to minds, private worlds were subjected to the national surveillance of fear. Addressing the problem of private fears, the Bush administration offered "freedom from fear" as an external system of total safety. This expansion of authority required its citizens to be patient during the unbounded war on terrorism, in a condition of waiting and obedience. Citizens were not asked to actively do anything, but rather to rid themselves and their families of fear: "Americans are asking: What is expected of us? I ask you to live your lives, and hug your children. I know many citizens have fears tonight, and I ask you to be calm and resolute, even in the face of a continuing threat"

(Bush, 2001, p. 762). This was very different than Roosevelt's vision in which freedom from fear was connected to active programs of social welfare and human rights. Roosevelt linked freedom from fear with global arms reduction, and with protection of people not only from external attacks but also from violence and oppression within their own nations. The newly formed United Nations adopted freedom from fear as an inherent right, and in its International Covenant on Economic, Social, and Cultural Rights (1966) explained that such an ideal would arrive only when "conditions are created whereby everyone may enjoy his economic, social and cultural rights, as well as his civil and political rights." Continually reaffirmed as a basic tenet of international law, this expansive combination of human rights and global responsibility was lost in a new paralyzing claustrophobia of inner surveillance.

REDEMPTION THROUGH CRISIS MANAGEMENT

Through deeply felt nationalism and desire for security, the administration shaped its crisis management, shoring up its threatened legitimacy. Crisis management is an attempt to restore faith in institutions shown by disaster to be insecure and flawed. Therefore, it works well with nationalism in redemptive narratives. As historians have noted, Judeo-Christian millennialism, which features purification and resurrection, often configures a disaster's imaginative dimensions, intersecting with patriotism, civic boosterism, and scapegoating (Biel, 2001; Dickie, Foot & Snowden, 2002; Johns, 1999; Linenthal, 2001; Smith, 1995; Steinberg, 2002). After disasters, self-interested institutions rush in to fix the sudden loss of meaning in normal life. They present renderings of disaster and recovery that best serve their own survivability, including evocations of a mythical bonding and renewed vigilance. The redemption of government after September 11 required not only that it provide emotional security and spiritual meaning, but also rational measures to eliminate threat. The result was an administration that evoked a spiritual authority meted out through purging of aliens, surveillance, and propaganda. Every citizen was expected to listen and watch for signs of evil in the community, for signs in the self.

In Ulrich Beck's (1992) now classic formulation, the risk society manufactures uncertain, uncontrollable hazards in a self-reflexive way, creating risk awareness to manage these hazards (p. 17; also Beck, 1992, 1999; Beck, Giddens & Lash, 1994; Giddens, 1991; Lupton, 1999). The same "modern reflexivity" operated in the post-September 11 national climate with its intense public examination of fear. Information collection and other invasive technologies contributed to the erosion of emotional privacy. Excessive measures were taken to make the citizen transparent to the state. In the empire of fear, the fantasy was total control, as citizens became willing subjects of the security state that both frightened and protected them. The Bush administration offered a pure "freedom from fear" as the inherent right of all citizens, an impossible goal that allowed unprecedented means, such as the USA. Patriot Act. It created and encouraged continual fears of amorphous unknown enemies, not just Al Qaeda operatives, in order to justify its ends and means. And it encouraged narratives of overcoming that fear as the highest patriotism. This rhetoric dominated the imaginative dimensions of the September 11 attacks and had direct consequences for people's lives, in the U.S. and abroad.

In many ways, the administration responded from the institutional lessons of crisis management, which it sanctified through religion and patriotism. The aim of crisis management is to protect corporate image, usually presenting the organization as a rational entity against the irrational forces of victims and critics. The institution is restored as vigilant and shielded, perfected by its ordeal. Crisis management seeks to narrow communications to repetition of simple optimistic messages, avoid debate with outsiders, pretend apology and remorse while denying responsibility, scale down and isolate operations, scapegoat and jettison potentially hostile workers, increase surveillance over remaining workers, and use internal public relations to ensure their absolute loyalty. Employee's emotions of grief, sadness, and fear are channeled into public relations campaigns to offset images of corporate heartlessness. I have elsewhere called this crisis tendency to narrow the organization while it asserts discursive control over its members, "lifeboat ethics" (Larabee, 2000, pp. 16–17). In the microcosm of the lifeboat, the survivors accept and even welcome strict regimes of triage, bodily control, and rationing because a deadly environment surrounds them. Therefore, persuading people that they are in such a lifeboat is an important crisis management strategy, emphasizing their dependence on the organization. The complicated, expensive work of compensating victims and rebuilding safe, sustainable communities is sacrificed to the drama of corporate preservation. After September 11, the reflexes of crisis management put national preservation to the fore, as the administration sought to narrow and streamline its activities, put strict controls over communications, arm all borders, cast out dissenters and aliens, and promote an emotionally resonant image of restored control.

CIVIL DEFENSE

Nostalgic, spiritualized post-disaster national preservation has resulted in a highly militarized security state that offers total safety through an internal Office of

Homeland Security, formed to protect the people against "invisible enemies" and "unknown threats of the future" (Bush, Homeland, 2002). A return to the old social forms of the cold war, civic preparedness is put forward in the less controversial language of community safety. Safer homes, safer communities, safer workplaces, healthier citizens – these are all the supportable goals of civic life. But as in the cold war, when even the domestic sphere was hardened against nuclear catastrophe, civil preparedness is directed against an ever-present external danger.

While defense against nuclear war was always an absurdity, civil preparedness against terrorism has an aura of authenticity. There are reasonable protections against chemical, biological, and limited radiation weapons. But the range of possible, unexpected weapons seems infinite and unpredictable, and so the terrain of fear is broadly extended to everyday objects. Like the shadows on the wall of Rockwell's painting, the very fabric of the ordinary world is transformed by terrorism and the war on terrorism. As Homeland Security Director Tom Ridge (2002) explained, "Homeland security is not an inside-the-beltway story. It encompasses the air we breathe, the food we eat, the water we drink, the energy we use, critical infrastructure everywhere. It affects us every time we board a plane or visit the office or log onto our computers. It touches everyone's lives."

Inhabitants of highly technological societies live among objects – like exploding coffeepots and E. coli-tainted hamburgers – that frequently endanger them, injuring hundreds of thousands every year. They make tacit psychological bargains with these threats that recede into the background of everyday life. They get into their cars and drive, even though there are 40,000 highway deaths a year in the U.S. Daily they face dangers caused by corporate and governmental negligence. And they often think of their relationship to sex, food, work, housing, and transportation as a process of assessing and managing personal risks. Risk is popularly understood as a mathematical probability that can provide relief from the groundless uncertainty of fear. In order to have economic and technological progress, so the narrative goes, businesses and research institutions must properly quantify risks and develop rational means for handling them. Occasional disasters should provide lessons, not lead to paralyzing fear. Even on the personal, psychological level, an ordinary person can assess risks before taking them, overcoming needless fear to make rational choices about driving, skydiving, or eating mercury-tainted tuna.

The discourse on risk has shifted significantly as danger seems to escape any control by the citizen. Fear has achieved cultural legitimacy over risk, as even the paranoiac's contamination nightmares are realized in the discovery of anthrax spores in the daily mail. The language of risk, which devalued fear, is too chilly for the national climate. Just as military strategists have understood that demonstrations of technological power can demoralize enemy soldiers and civilian populations, terrorist groups use limited resources to achieve mass psychological effects. With global news media, deep fear and sorrow can touch viewers who are very remote from the scene of violence. Managing these feelings, channeling them into wartime service through public persuasion, becomes the task of the modern state experiencing such attacks. Fears have risen to occupy a central place in public discourse, as the relationship between citizens and the new security state is negotiated. Whether the danger is real or imagined is unknowable and irrelevant; there is no distinction between fear, which in psychological terms is always attached to an object, and general objectless anxiety. It follows that the greater the fear, the greater the patriotic sacrifice. Even if the citizen's decision to fly is not based on patriotism, it is subsumed into a larger narrative of America's choice, as all citizens are presumed fearful of the threat to the nation. As Attorney General John Ashcroft (2002) put it, "all innocent civilians are exposed to the threat of wanton violence and incapacitated by the fear of terror." A citizen's decision to take a plane to New York is no longer a private matter of overcoming her fear of flying by reading statistical probabilities of a crash. Nor is it important whether media spectacles of plane crashes caused her fear. What matters is that the citizen braves her fear for the sake of national honor and survival.

Thus, the logic of the war on terrorism transforms even daily living into a politically significant act. A few weeks after the September 11 attacks, President Bush told citizens to "get on board," to live their lives with the larger purpose of supporting America's businesses. Getting out and working, going on vacations, buying consumer goods all have become part of the national goal. Furthermore, citizens are asked to engage in surveillance of their friends, neighbors, families, households and communities. The Federal Emergency Management Agency (2002) suggests that citizens learn about terrorism and be always on the alert for suspicious persons and packages, and for "heavy or breakable objects that could move, fall or break in an explosion." The National Crime Prevention Council's *Citizens' Preparedness Guide* (2002) urges patriotic citizens to be alert, to study the routines in their neighborhoods and note "anything out of place." Citizens should get to know their neighbors, not to generate warmer communities, but to be "on the lookout for suspicious activities such as unusual conduct" (p. 1). As in the Cold War days of nuclear preparedness, citizens are supposed to imagine worst-case scenarios, and plan accordingly, keeping supplies of food and water on hand. Always aware of a threatening, deliberately violent agent lurking in the everyday environment, the fearful subject is the perfect citizen.

The citizen is encouraged to monitor her community for "unusual conduct" or "anything out of place." And, in a culture where bodily self-surveillance is a mania, she is encouraged to monitor her physical and mental being for the good of the nation. Thus, when President Bush (2002) launched a fitness program, including admonitions to eat healthy and quit smoking, he said, "if you're interested in improving America, you can do so by taking care of your own body" (Health). And, speaking of his health initiative, he told a group of senior citizens in Florida, "[The terrorists] must have thought we were weak. They must have thought we were really a feeble country. But they're learning differently. You see, when it comes to the defense of freedom, we're strong and we're plenty tough" (Physical Fitness, 2002). Caught up in social ideas about deviant and acceptable bodies, dietary and fitness regiments are nevertheless popularly considered personal choices, taking control over one's risk of loneliness or disease. Now the body is a sign of national strength. Even eighty-year-old senior citizens are enlisted to imagine their bodies as tough enough to fight off terrorists. In a world of invisible enemies, internal monitoring and surveillance of neighborhoods prevents an invasion of the body snatchers, a national paranoia not seen on this scale since the 1950's.

THE INSECURITY STATE

By promising total security, citadel culture can easily deliver insecurity. When the war turns within, the fight against subversive internal energies is both reassuring and worrying. The discourse of the internal war on terror is often uneasy, overwrought, and contradictory, since total control is simply impossible. No seamless narratives of a pure nation can prevent the ontological insecurities of death. As it seeks the deepest control over its citizens, the apparatus of the security state, steeped in its own chronic fear and panic, turns in on the people it seeks to protect.

The mysterious, unsolved, internal threat of the anthrax scare provides an example of an overwrought reaction to invasive agents, fragile systems, and vulnerable bodies. In the midst of its redemptive narratives of healing and national recovery, the national government and the news media were stricken with panic over spores found in the daily mail. Frightened for their own lives in the first days of the anthrax scare, employees at national news agencies were intensely interested in any rumor of it, not necessarily in the national interest. During this period of fear over viral agents, CNN broadcast the news that an anthrax incident had been reported at Michigan State University.

Thus, a strange scene unfolded on my university campus. In the building that houses Michigan State's College of Arts and Letters and the Graduate Office, an office worker called the campus police to report that after she opened a letter, she felt a burning sensation in her throat. Rumors were started that an envelope with white powder had been found in the building, and that animal rights activists, who had previously started fires on campus, were using the climate of fear to promote their cause. There was, however, no animal research anywhere in or around the building. And there was no initial attempt to verify that the report and the rumors were true. The FBI, four police departments, a hazardous materials team and an ambulance unit were immediately sent in. Fifteen women were evacuated, and then hosed down and scrubbed with bleach. Especially humiliating, they were decontaminated in an open room by a police emergency staff consisting of all men. One of the women who resisted was threatened with force. What transpired during the decontamination was not known to most of us in the campus community until the women, still feeling traumatized a year later, agreed to speak at a local American Civil Liberties Union meeting (Englund, 2002, p. 1).

In this incident, the new apparatus of fear, alertness, preparedness, and institutional response in the war against terror only ended up assaulting and humiliating a few working women. However, for CNN, which was in some ways responsible for the national hysteria, the panic and insecurity of the anthrax scare were immediately subsumed in the larger linear narrative of resuming control and winning. In the words of CNN medical advisor Sanjay Gupta, "We proved to ourselves that this infection was something that we could get a handle on, we could treat people . . . we beat that" (Cable News Network, 2003). The pressures of imagined community, of national transcendence, erased local worlds where negotiations between people and the new security apparatus were painfully, chaotically taking place.

Though these regimes of security would not ordinarily be tolerated in a democracy, they have inspired faith with their promise of technological control over borders, bodies, and information. Total Information Awareness offers to make all citizens completely transparent to the state, eliminating any hint of fear. Risk is not to be proportioned and managed, but experienced as the price of imagined community. Thus, the state, more than any postmodern risk theorist, understands that risk is constructed through subjective perception and social, political, and historical processes. Perhaps the most skilled "postmodernists" reside in these political institutions orchestrating the war against terror at home. The spontaneous emotional responses to September 11, unfolding as modalities of understanding, were publicly reduced and shaped into fear in the service of imagined community. This project is much larger than the construction of a single risk or a hierarchy of risk, for it is a deliberate project of social transformation in which fears are attached to anything out of order. Risks are understood as the subjective perception of the "unusual" in persons and things. And all fearful citizens become soldiers in the war on terror, equally subject to secret investigation based on ungrounded suspicion. Although moral panics, red scares, and wartime hysteria always fade, the residue of an oppressive state apparatus can remain. Therefore, it is worth returning to the cool registers of the risk society, restoring a critical self-reflection to the post-disaster and wartime discourse on fear and risk.

REFERENCES

Ad Council (2002). September 11 [Online]. Retrieved October 12, 2002, from http://www.adcouncil.org/issues/September_11th/

American Council of Trustees and Alumni (2002). *Defending civilization: How our universities are failing America and what can be done about it* [Online]. Retrieved July 21, 2003, from http://www.goacta.org/publications/reports.html

Anderson, B. (1991). *Imagined communities: Reflections on the origin and spread of nationalism*. New York: Verso.

Ashcroft, J. (2001). Excerpts from Attorney General's testimony before Senate Judiciary Committee. *New York Times*, December 7, p. B6.

Ashcroft, J. (2002). National security: Justice Department's new mission. *Vital Speeches of the Day, 68*, 706–708.

Beck, U. (1992). *Risk society: Towards a new modernity*. Mark Ritter (Trans.). London: Sage.

Beck, U. (1999). *World risk society*. Cambridge: Polity.

Beck, U., Giddens, A., & Lash, S. (Eds) (1994). *Reflexive modernization: Politics, tradition and aesthetics in the modern social order*. Cambridge: Polity.

Biel, S. (2001). *American disasters*. New York: New York University Press.

Bush, G. W. (2001). We are at war against terrorism: Address to Congress. *Vital Speeches of the Day, 67*, 760–763.

Bush, G. W. (2002a). Homeland Security Proposal, June 18 [Online]. Retrieved September 7, 2002, from http://www.whitehouse.gov

Bush, G. W. (2002b). Press release. President outlines health and fitness initiative, June 20 [Online]. Retrieved September 7, 2002, from http://www.whitehouse.gov

Bush, G. W. (2002). President promotes physical fitness to senior citizens, June 21 [Online]. Retrieved September 9, 2002, from http://www.whitehouse.gov

Bush, G. W. (2002). Securing freedom's triumph. *New York Times*, September 11 [Online]. Retrieved January 5, 2003, from http://www.nyt.com

Cable News Network (2003). *America remembers*. AOL Time Warner.

Cooper, G. (2002). Darts and daggers. *Columbia Journalism Review*. September–October [Online]. Retrieved September 9, 2002, from http://www.cjr.org/year/02/1/dartsandlaurels.asp

Dickie, J., Foot, J., & Snowden, F. M. (Eds) (2002). *Disastro! disasters in Italy since 1860: Culture, politics, society*. New York: Palgrave.

Englund, E. (2002). A call for dignity: Workers recall anthrax scare. *State News* (Michigan State University), October 14, p. 1.

Federal Emergency Management Agency (2002). *Terrorism fact sheet* [Online]. Retrieved September 27, 2002, from http://www.fema.gov

Giddens, A. (1991). *Modernity and self-identity: Self and society in the later modern age*. Cambridge: Polity.

Giuliani, R. W. (2001). Eradication of terrorism: A matter of justice leading to peace. *Vital Speeches of the Day, 68*(1), 3–6.

Hoffman, B. (1999). *Inside terrorism*. New York: Columbia University Press.

Johns, A. (Ed.) (1999). *Dreadful visitations: Confronting natural catastrophe in the age of enlightenment*. New York: Routledge.

Larabee, A. (2000). *Decade of disaster*. Urbana: University of Illinois Press.

Linenthal, E. T. (2001). *The unfinished bombing: Oklahoma City in American memory*. Oxford: Oxford University Press.

Lupton, D. (1999). *Risk*. New York: Routledge.
National Crime Prevention Council (2002). *United for a stronger America: Citizens' preparedness guide*. Washington, DC: NCPC.
National Public Radio (2002). Transcript. Attempts by the Ad Council to sell Americans on the need to guard their freedom closely. Morning Edition, August 26.
Parker, R. (2002). Homeland: An essay on patriotism. *Harvard Journal of Law and Public Policy, 25*, 407–427.
Ridge, T. (2002). Remarks to the National Association of Broadcasters Education Foundation 2002 service to America Summit, June 10. Retrieved October 21, 2002, from http://www.whitehouse.gov
Smith, C. (1995). *Urban disorder and the shape of belief: The great Chicago fire, the Haymarket bomb, and the model town of Pullman*. Chicago: University of Chicago Press.
Steinberg, T. (2002). *Acts of God: The unnatural history of natural disaster in America*. New York: Oxford University Press.
Teinowitz, I. (2002). Selling the USA. *Advertising Age*, July 1, 3, 49.
United Nations (1966). International covenant on economic, social, and cultural rights, December 16 [Online]. Retrieved April 8, 2003, from http://www.un.org/millennium/law/iv-3.htm

DISASTER BELIEFS AND INSTITUTIONAL INTERESTS: RECYCLING DISASTER MYTHS IN THE AFTERMATH OF 9–11

Kathleen Tierney

ABSTRACT

Research evidence developed over more than five decades of research on human responses to disasters shows that those responses are overwhelmingly adaptive and positive. However, despite what is known, myths about disaster behavior persist. These include the assumption that the public will panic during large-scale emergencies and the idea that disasters are best managed through hierarchies of command and control. Following the tragic events of September 11, 2001, these myths are again gaining wide currency even though actual individual, group, and organizational behavior in the World Trade Center disaster directly contradict those assumptions. This is no accident. Beliefs concerning the fragility of the public in the face of emerging homeland security threats are consistent with the perspectives and objectives of organizations that seek to expand their influence in the domestic crisis management arena. These organizational actors, which include the information technology industry, the intelligence and defense establishment, and security think tanks, are generally not familiar with empirical social science research on behavior during disasters and see little value in public participation in the management of newly-recognized threats. Recycled

Terrorism and Disaster: New Threats, New Ideas
Research in Social Problems and Public Policy, Volume 11, 33–51

ISSN: 0196-1152/doi:10.1016/S0196-1152(03)11004-6

disaster myths support a case for "expertise-based" crisis planning that excludes the public from those activities.

INTRODUCTION

Thirty years ago, in February 1972, an article by sociologists Henry Quarantelli and Russell Dynes appeared in *Psychology Today* (Quarantelli & Dynes, 1972). That article, which was entitled "When Disaster Strikes (It Isn't Much Like What You've Heard and Read About)," made the point that a great many of the things people believe about social behavior in disasters are simply contrary to the findings of social science research. The article went on to discuss a number of myths or erroneous images of disaster behavior. Those myths included the notion that panic is widespread in disaster situations; the idea that disasters lead to collective demoralization and social disorganization, which then provide a context for the emergence of anti-social behavior, such as looting; and the assumption that, rather than carrying out their designated duties when disasters strike, emergency workers will abandon their posts in favor of saving their loved ones. Using data obtained through pioneering field studies that had been conducted by the National Opinion Research Center, the National Academy of Sciences, and the Disaster Research Center during the 1950s and 1960s, the article rebutted these myths and replaced them with much more positive, adaptive images of social behavior in disasters.

Since that *Psychology Today* article appeared, more intensive research on disasters in the U.S., as well as research in other societies, has shown that many of the findings from early disaster studies are very robust. At the same time, however, other research has indicated that mythical thinking about social behavior in disasters persists. Various studies have shown that despite evidence to the contrary, the general public, governmental officials, and even many disaster victims themselves continue to think about disaster-related behavior in terms of erroneous images (Fischer, 1998; Wenger et al., 1975; Wenger, James & Faupel, 1985). Disaster scholars argue that these views have likely been influenced by the ways in which disasters are framed in the mass media and in popular culture (Clarke, 2002; Mitchell et al., 2000; Quarantelli, 1985). However, less attention has been paid to the ways in which disaster myths serve to bolster the claims of institutional actors and organizations that manage risks.

This latter aspect of disaster mythology – the idea that myths concerning disaster behavior and disaster management reinforce particular institutional interests – is the topic of this analysis. I begin by noting that following the attacks on the World Trade Center and the Pentagon and the anthrax attacks that took place in fall, 2001, old myths about the manner in which people and organizations respond under conditions of extreme danger are being recycled, both in public discourse and in governmental planning for emerging threats. My discussions focus on two of these recycled myths: the panic myth and the notion that the way to manage disasters is through hierarchies and centralized command-and-control structures. I then contrast these mythical ideas with findings from empirical research on disaster events, including the World Trade Center attack. The final section of the paper discusses how beliefs concerning the fragility of the public in the face of emerging threats are consistent with the perspectives and objectives of organizations that have been seeking to expand their influence in the domestic crisis management arena in the wake of the tragic events of September 11.

PANICKY VICTIMS AND THE NEED FOR CONTROL: MYTHICAL CLAIMS VERSUS SOCIAL SCIENCE RESEARCH

A recent review by Clarke (2002) starts by noting the important role that panic serves as a plot device in disaster movies and moves on to discuss how assumptions about the prevalence of public panic under crisis conditions have influenced recent emergency preparedness activities, such as governmental planning for Year 2000 computer failures. Clarke juxtaposes mythical images of panic against actual data on the adaptive and prosocial manner in which individuals and groups have behaved in crises ranging from deadly conflagrations to plane crashes. The points he makes have been borne out again and again in disaster research: despite the fact that people may well be terrified in disaster situations, even to the point of feeling that their lives are in imminent danger, they almost never resort to the kind of highly individualistic, competitive, headlong flight behavior that characterizes true panic. Instead, social bonds remain intact and the sense of responsibility to others – to family members, friends, fellow workers, neighbors, and even total strangers – remains strong (Johnson, 1987; Johnson, Feinberg & Johnson, 1994; Johnston & Johnson, 1989).

The same patterns of rational, orderly response and helping behavior that have been observed in other disasters were evident in numerous first-person accounts on the evacuation of the World Trade Center. Despite media reports that continually described occupants of the Trade Towers and evacuees as panicky, what was seen instead were people behaving with admirable presence of mind under the most adverse of circumstances. Endangered tower occupants sought information from one another, made inquiries and spoke with loved ones via cell phones, engaged in collective decision making when they encountered danger, and helped one another to safety. When occupants escaped from the towers, the evacuation, which was highly successful by any standard, was carried out in a calm and orderly manner (see, for example, the detailed account of evacuee decision-making compiled by Cauchon, 2001, for USA Today). Similar patterns of highly adaptive and pro-social behavior were documented following the bombing of the World Trade Center in 1993 (Aguirre, Vigo & Wenger, 1998).

The incidence of panic behavior is vanishingly rare in actual crisis situations, even those that are life-threatening. Of course, this does not mean that people feel no fear when disaster strikes, or that they do not try to escape from danger. The point is that despite fear and the entirely realistic desire to flee, individuals and groups facing severe danger try to make rational decisions, and they maintain their connections with others. For example, researchers have shown that in a catastrophic nightclub fire that killed nearly two hundred people, there were a few isolated cases of panic, but on the whole those who were trapped and searching for ways out of the club as the fire raged behaved in an orderly and mutually supportive fashion (Johnson, Feinberg & Johnston, 1994; Johnston & Johnson, 1989).

Influenced by strong norms of altruism that emerge during major emergencies, those affected respond in ways that are positive and constructive. The literature on disasters consistently shows that far from fleeing in terror when disasters strike, disaster "victims" actually constitute the real first responders in those situations. Numerous studies document the fact that individuals and groups in the immediate impact area manage evacuations, perform rescues, locate and dig out fellow victims who are trapped, transport them to emergency care providers, and put themselves in danger repeatedly to ensure that others are safe (Aguirre et al., 1995; Dynes, Quarantelli & Wenger, 1990; Noji, 1989; Tierney, 1994). Media reports on disasters such as major earthquakes and on terrorist incidents like the attack on the Murrah Federal Building in Oklahoma City and the World Trade Center tend to feature stories on the rescue activities of official governmental search and rescue teams, leaving community residents and other unofficial rescuers out of the picture. The nation has twenty-eight such teams around the country, consisting of trained personnel, heavy equipment, and search dogs. What the media do not state is that while those teams do search, they virtually never rescue. Instead, the vast majority of live rescues are carried out by community residents who are at the scene of disasters, not by official response agencies or outside search and rescue teams.[1] Indeed, rather than panicking and acting in otherwise unproductive ways, members of the public typically converge en masse to help when disasters strike. Behavior following the World Trade Center attack was no different. Victims helped fellow victims to safety, and the wave of people leaving Lower Manhattan when the Towers collapsed was accompanied by a counterwave of those converging into the area to help.

Despite extensive empirical evidence to the contrary, many crisis planners appear to assume that in any future terrorist attack, panicky community residents will be a major problem and perhaps even a danger to public order. Much current U.S. bioterrorism planning is predicated on the assumption that the public will panic and flee in horror should such an attack occur. Writing about the U.S. Centers for Disease Control plans for bioterrorism in this country, a Washington Post article observed that "The CDC plan makes it clear that if an attack occurred in the United States, avoiding mass panic would be a major challenge" (Washington Post, November 27, 2001). Indeed, one prototype CDC bioterrorism plan, developed prior to 9–11, states that "Following a terrorism-related event, fear and panic can be expected from both patients and health care providers" (Centers for Disease Control and Prevention, 1999). In the May, 2000 TOPOFF crisis management exercise, which involved weapons of mass destruction, high-ranking officials taking part in the exercise conjured up images of mass public panic, which then had to be managed. In the fall of 2001, the entirely understandable and predictable information-seeking among the public that occurred in connection with the anthrax attacks was labeled panic by many U.S. media.

After attending a meeting held by the Royal Society and the British Association for the Advancement of Science in December, 2001, an observer posted a report on the Internet noting that "One certainty in this very uncertain area is that panic may cause more havoc than any biological agent . . . the English Department of Health has struck the correct balance with well prepared scenarios communicated to officials and practised from time to time while downplaying the information to the public to avoid panic" (Symons, 2001). Recent newspaper reports (see Gellman, 2002) indicate that similar thinking dominates U.S. bioterrorism planning. As in Great Britain, concern with creating panic is being used as a rationale for not communicating with the public or including the public more directly in counter-terrorism planning. In one notable case that came to light in spring, 2002, the federal government avoided informing public officials in New York City about a credible threat involving nuclear material that had surfaced in fall, 2001, ostensibly to avoid creating panic. Subsequent media coverage again promulgated the panic myth, as illustrated in this transcript of an interview conducted on CNBC with Bernard Kerik, former New York City Police Commissioner by television personality Chris Matthews (CNBC, Inc., 2002):

> Matthews: What would you have done with that information [the warning concerning nuclear terrorism] without letting it out of the bag? . . .
>
> Kerik: . . . I can't tell you exactly what we would do in New York City . . . but what I will tell you is that we have 41,000 New York City police officers, and if there was a need to conduct a search . . . I think we would have been . . . much better prepared by notifying the police department and the appropriate authorities in New York. . . .

Matthews: How would you have described the – how would you have described the targeting of a search without causing the most historic, almost Biblical level of panic in Manhattan?... Suppose a person in New York, who's living in Manhattan, hears there may be a nuclear weapon blowing up in New York. They would grab their kids. They'd begin walking to the nearest bridge if they had to. They would have taken a subway if they were still running. You would have almost a Japanese kind of subway craziness going on, wouldn't you?

Kerik: Well, I think that you have to walk a fine line on what you – what you let out and what you do not.

Japanese subways are well-known for the orderly and civil manner in which passengers comport themselves, particularly during crowded rush hours, and moving toward areas of safety is clearly a rational response to danger. Nevertheless, the media remain undaunted in their search for panic. For example, in a virtual tour de force of disaster mythology, an article published on the Web by the Newhouse News Service (Wood, 2002), entitled "America Is Dangerously Vulnerable to Panic in Terror Attack, Experts Say," speaks grimly of the "widespread, blind panic" that could break out among the U.S. population in the event of repeated terrorist attacks – panic episodes that will be accompanied by looting and civil unrest. An expert source is quoted as being "worried about terrorism causing the collapse of civil society," while elsewhere the article discusses the need for law enforcement crackdowns in bioterrorism emergencies in which "[p]olice barricade city and state borders. Armed guards patrol quarantine lines that may divide neighborhoods and even families."

Beliefs about the specter of public panic in future terrorist attacks are widespread despite published work by researchers like those associated with the Institute for Civil Biodefense Strategies at Johns Hopkins University, indicating that on the basis of past research they do not expect the public to panic even in the face of biological threats and that, on the contrary, the public should be seen as an important resource for responding to those kinds of attacks. Institute researcher Thomas Glass points out, for example, that in the massive 1918 flu pandemic, which killed 500,000 U.S. residents, "people responded quite effectively and implemented disease control and treatment strategies in the community... Little evidence of panic can be discerned from the historical record" (Biohazard News, 2001, p. 3). Similarly, accounts of the 1947 smallpox outbreak in New York City, which led to the very orderly vaccination of six million New York residents, make no mention of public panic (Weinstein, 1947). Glass has also spoken disapprovingly of what he calls "the yellow-tape effect" in terrorism planning – the tendency on the part of official crisis management agencies to be suspicious of the public and to want to keep volunteers at arms length because of their lack of qualifications and potential disruptiveness (Biohazard News, 2001). He and his colleagues have called instead for the provision of timely and accurate information to the public – not the withholding of information – as well as the inclusion of community residents into local preparedness efforts as the most effective way of countering any potential for panic that may exist (Glass & Schoch-Spana, 2002).

While concerns persist about not stirring up panic, there is little discussion on the very real challenges associated with emerging threats, such as how to communicate emergency information to the public in appropriate ways, and, should the need arise, how to issue warnings effectively in a society with pronounced social class, ethnic, and language divisions. Indeed, concern with keeping information from members of the public so that they will not panic may well obscure a more serious risk – the very real possibility that crucial information on hazards and recommended self-protective measures will not be equally available to all groups in our diverse society should it be needed following some future terrorist attack.

Hierarchies of Command: Militarizing Emergency Management

Paralleling the myth of panic and dysfunctional behavior on the part of the public is a growing emphasis on the need for stricter lines of command and control in disasters. Both sets of imagery are premised on the notion that extraordinary measures are needed to counteract the social disorganization and precursors to social breakdown that are thought to accompany disasters. Disaster scholars have long called attention to the existence in both political and scholarly discourse of two very different models of disaster response and management. The first, which is based on what actually occurs in disaster situations, is commonly termed the problem-solving or human resources model. The second, known as the military or command-and-control model, is based on the idea that the appropriate way to manage disasters and other crisis situations is through centralized control and clear-cut hierarchies (for discussions, see Dynes, 1993, 1994). This mythical command-and-control image persists in part because disasters are erroneously seen as situations in which public order breaks down, and in part because the field of disaster management itself has its roots in the Cold War and in military notions of civil defense.

The command analogy appears in various ways in discourses on disasters and their management. Following Hurricane Andrew in 1992 – a disaster in which the intergovernmental response was slow and judged inadequate by many – there were discussions on whether the military should be given a greater role in domestic disasters, since the military has the command structure and the know-how to get things done.[2] Critiques of the handling of the World Trade Center response by the

New York City fire and police departments (see McKinsey & Company, 2002a, b; New York Times, 2002a, b), have centered on the idea that no one appeared to be in charge of the overall response, that command structures broke down, and that front-line workers weren't sure to whom they were reporting on 9–11. Governmental plans for the creation of the new federal homeland security agency are based on the notion that hierarchical organizational arrangements are required to deal with emerging homeland security threats – even though organizational research suggests that hierarchies are likely to be especially ill-suited to managing those kinds of risks (Wise, 2002).

Those who argue for greater command and control capability and hierarchical structures for managing future crises fail to take into account how decisions are actually made and how activities are carried out in disaster situations. Disasters have long been characterized in the sociological literature as crisis occasions in which demands on affected social units exceed their capability to handle those demands. Faced with so many pressing problems, social units are forced to adapt. For example, in disasters, many decisions must be made under extreme time pressure. Research indicates that, due to such demands and time constraints, decision making typically devolves to lower levels in crisis-response organizations than would normally be the case, and front-line responders and mid-level personnel have greater autonomy to act than during non-disaster times (Drabek & Haas, 1969). In another common pattern, emergent groups – that is, social groupings that had no existence prior to the occurrence of the disaster – organize in order to deal with disaster-related problems that their members define as pressing (Stallings & Quarantelli, 1985). Group emergence occurs in a variety of contexts in disasters. In many cases, groups emerge and carry out their activities entirely outside the formal emergency response structure, as when community residents organized themselves to prepare and deliver meals to workers at Ground Zero following the World Trade Center attack. In another common pattern, new networks form that blend the activities of existing organizations with those of emergent groups. In an example of this type of blended network, building and safety officials worked with volunteer structural engineers from New York and from around the country to conduct safety inspections of hundreds of buildings in and around the Ground Zero.

Emergence is also accommodated within existing crisis response organizations, as activities performed by volunteers are incorporated into the official response. For example, following the Trade Center attack, volunteers who came primarily from local universities began staffing a geographic information systems (GIS) and mapping unit within New York City's emergency operations center. That unit, which expanded in the days and weeks after the attack, provided important information and support services to government officials and responding agencies.

Relatedly, and also directly contradicting top-down images of command and control, because disasters always contain elements of surprise, they always become occasions for extensive organizational improvisation. These improvisational activities come about not through centralized control and decision-making but through efforts to deal with local problems as they emerge, through the problem-solving activities of informal networks, and through processes of bricolage that use existing resources in novel ways. On the day of the World Trade Center attack, for example, in a story that was undercovered by the mass media, several hundred thousand people were evacuated from Lower Manhattan by water via an emergent network of private and publicly-owned watercraft – a previously unplanned activity that lacked central command and control. Those same vessels brought rescue workers, volunteers, and donated supplies to the impact area (Kendra & Wachtendorf, 2002; NYCStories.com, 2001; Snyder, 2001; Stravelli, 2001). The volunteer engineers who worked with City building officials to inspect damaged buildings at and around the Ground Zero site borrowed and adapted a rapid-damage-screening protocol that had been originally developed by engineers in California for conducting building safety assessments following earthquakes. They used the information that was collected to produce a database and maps indicating which buildings were safe and which were hazardous, in order to advise City officials and building owners. Those activities were improvised; New York City had no plan for rapidly screening damaged buildings prior to September 11. The multiorganizational network that developed to deal with the massive task of debris removal from the World Trade Center, while made up of existing organizations, was developed after the attack during the course of the emergency response itself. The organizational roles and relationships that crystallized within that network were very different from those envisioned in prior planning by the City. The stabilization of the structures at the Ground Zero site and initiation of debris removal were key elements in the emergency response to the Trade Center attacks. On September 11, the City's Department of Design and Construction acted rapidly to define a role for itself in the coordination of those activities – a disaster role for which that agency had no prior authority (Langewiesche, 2002). The use of geographic information systems, computer modeling, remote sensing, and other techniques and technologies to deal with numerous other problems as they developed following the terrorist attacks was almost entirely improvised. And in what may have been the most striking example of improvisation of all, the Mayor's Office of Emergency Management was able to completely reconstitute its emergency operations center because its state-of-the-art facility, which had been located at 7 World Trade Center, was completely destroyed on September 11.[3]

The lesson here is that the response to the September 11 tragedy was so effective precisely because it was *not* centrally directed and controlled. Instead, it was flexible, adaptive, and focused on handling problems locally as they emerged. Response activities initially involved mainly those who were present in the immediate area when the attacks occurred and that later merged the efforts of officially-designated disaster response agencies with those of newly-formed groups, as well as literally thousands of other organized entities that had not been included at all in prior emergency planning and that were not subject to any central authority.

Actual community crisis response networks look nothing like the command-and-control hierarchies that are the stuff of disaster myths. Rather, they consist of loosely-coupled collections of individuals, groups, and organizations that continually change and that have permeable boundaries. Command-and-control models do not work in disasters because authority and responsibility are distributed, not centralized, and because so many of those who take part do so outside the boundaries of official agencies. Some dimensions or phases of disaster response activities, such as firefighting, can and do involve clear-cut chains of command, but the term that more appropriately describes the overall response to disasters is co-ordination – and in some cases, organized anarchy – rather than control.

Rather than being organized according to principles of command and control, disaster response activities are undertaken through a complex and varied set of organizational arrangements characterized by a high degree of emergence and improvisation. Some scholars characterize disaster responses as multiorganizational garbage cans (Beamish, 2002; Clarke, 1989) in which solutions and problems merge in ways that are unexpected and unplanned. Others view emergent disaster response networks as complex adaptive systems (Comfort, 1999). Still other analysts argue that network management is an appropriate term for the manner in which disaster response activities are coordinated (Waugh & Sylves, 2002; Wise, 2002). Indeed, the prevalence of network forms of organization (Podolny & Page, 1998), as opposed to hierarchies, is a distinguishing feature of the organized response to major disaster events.

Instead of indicating weakness or management flaws, the decentralized multiorganizational responses that followed the World Trade Center attacks and that have repeatedly been documented in major disasters are in fact a major strength and a source of community and societal resilience. Centralization and hierarchy slow down and otherwise hamper response efforts. Their flaws were highlighted following a major earthquake that struck the Kansai region in Japan in January, 1995. That disaster, which affected the city of Kobe and nearby communities, resulted in the deaths of more than 6,000 people and left nearly 300,000 people (approximately 20% of the population of the impact region) homeless. The initial governmental response to the earthquake was widely viewed as both sluggish and ineffective. The inability to move expeditiously to address the very severe problems that resulted from the earthquake stemmed in part from the highly centralized and hierarchical nature of the Japanese governmental system – a system that was not flexible enough to allow for autonomous action on the part of local agencies. On a scale never before documented in a disaster in Japan, the earthquake triggered a massive convergence of volunteers and the emergence of numerous citizen groups in the disaster-stricken region. This volunteer mobilization, which involved an estimated 630,000 to 1.3 million people (International Federation of Red Cross and Red Crescent Societies, 1996), helped compensate for the government's lack of foresight and for gaps in response capability. However, many local jurisdictions had great difficulty incorporating volunteer groups into their response and recovery efforts, again owing to their unwillingness to permit deviations from standard procedures and formal lines of authority, even under the most extreme disaster conditions (Atsumi et al., 1996; for an overview of the 1995 earthquake disaster, see United Nations Centre for Regional Development, 1995).

Following the 9–11 attack, scholars have questioned what they see as a movement toward more militaristic, command-and-control approaches to the management of the terrorist threat. They even argue that the adoption of such models may well make the nation less safe in future attacks. For example, Waugh and Sylves (2002) stress the notion that responses to disasters are characterized by complex multiorganizational arrangements and a variety of forms of interorganizational and intergroup collaboration and coordination, ranging on a continuum from formal contracts to informal relationships based on trust. They warn that given these characteristics of emergency response networks, it would be a major mistake to (2002, p. 83):

> superimpose overbearing command systems riddled with secrecy requirements that complicate the collaboration and public involvement essential to dealing with hazards and disasters . . . [I]f the war on terrorism inadvertently undercuts or distorts an emergency system designed to deal with so-called routine disasters, it may well weaken current capabilities to manage conventional hazards *and* the hazard posed by terrorism

Commenting on plans for organizing federal homeland security efforts, Wise singles out hierarchical structures as especially inappropriate for managing the war on terrorism, particularly in light of the environment in which organizations will be carrying out their activities (2002, p. 132):

> Adoption of standard, rational, hierarchical designs and practices is likely to be particularly unsuitable for organizations that are expected to operate in complex, unstable environments . . . More unstable environments create a need for greater decentralization of authority and less emphasis on formal structure, because a shifting environment requires rapid decisions and changes, and it takes too long for information and decisions to travel up and down a strict hierarchy.

In sum, of all possible ways to characterize responses to disasters – and also to manage those events – images of command-and-control and hierarchical organizational structures may well be the least appropriate. Such organizing principles are inconsistent with what research indicates actually occurs in disasters and ill-suited to the complex problems and challenges disasters present.

WHY NOW? THE APPEAL OF RECYCLED MYTHS

When September 11 offered so many examples of public heroism and selfless devotion to others, why is there now such a focus on the specter of public panic? When September 11 demonstrated the enormous resilience of our civil society, why is disaster response now being characterized in militaristic terms? Media influences are clearly relevant here, but recycled myths and discourses of panic and command and control are also a reflection of two other and equally important – or perhaps more important – influences: first, a movement on the part of governmental agencies and private-sector organizations, particularly those associated with defense, intelligence, and law enforcement, to claim ownership over the management of emerging threats; and second and relatedly, the pressing of claims regarding the need for management of future crises by those with particular kinds of specialized expertise.

The history of crisis management in the Federal Emergency Management Agency and its predecessor agencies has been accompanied by tensions between militaristic or civil defense approaches, on the one hand, and an emphasis on community-based preparedness for natural and technological disasters, on the other (Kreps, 1990). While the latter management orientation – that is, the notion that the U.S. emergency management system should focus on disasters and include a wide spectrum of community institutions – had become more prominent during the 1980s and 1990s, that approach never truly supplanted the war-oriented command and control model of management within FEMA and other federal agencies. In the wake of 9–11, this persistent institutional ambivalence is once again being resolved in favor of a more militarized approach to disasters and human-induced threats. New institutional actors had become interested in disaster management even before the September 11, 2001 terrorist attacks. Since the end of the Cold War in the early 1990s, the defense and intelligence sectors of the federal government had been seeking a larger role in U.S. disaster management. This movement was evident in initiatives such as the Global Disaster Information Network (GDIN), which attempted to employ technologies originally developed for military intelligence to hazard assessment and emergency response activities in natural disasters (Disaster Information Task Force, 1997).[4] National Laboratories such as Los

Alamos National Labs, as well as consulting companies that frequently work with the defense establishment, were also becoming increasingly involved in the natural hazards area.

The events of 9–11 have opened a major window of opportunity for those types of organizations and have also resulted in previously unimaginable levels of funding for crisis preparedness, training, and response – and in particular for the development and transfer of technologies that support command and control functions. Whatever their strengths with respect to technology and hardware – and those strengths are clearly considerable – the types of organizations that are seeking to expand their domains in the aftermath of 9–11 generally have little background either in understanding social behavior in disasters or, in many cases, in understanding how disasters are actually managed in U.S. society.

New federal homeland security initiatives launched in response to the 9–11 attacks signal a growing emphasis on top-down, command-and-control approaches to managing future crises. This shift is evident in the composition of newly-formed advisory committees, new agency alignments, and federal funding priorities. Some members of the sixteen-person Homeland Security Advisory Council appointed by President George W. Bush in June, 2002 have ties to local communities and a broad understanding of the nation's disaster management system. The mayor of Washington, D.C. is a member of the Council, for example. Another member has an extensive background in disaster management, having served as the director of emergency services in a disaster-prone state. By and large, however, the Council consists of individuals with highly technical backgrounds and training whose careers have been spent in such organizations as the CIA, the FBI, MITRE Corporation, Mitretek Systems, ANSER Institute, and other entities with close ties to the intelligence and defense establishments. The Federal Emergency Management Agency has now been absorbed in its entirety by the Department of Homeland Security, where its personnel are likely to play a relatively minor role compared with officials from other more powerful defense and security-oriented agencies. FEMA's organizational culture, never entirely comfortable with broad community involvement in emergency preparedness and response, will undoubtedly shift toward a greater emphasis command and control in all types of emergencies.

Federal budget allocations for homeland security are moving the emergency management system into a command and control mode, directing funds primarily to border and transportation security, IT-related activities at the federal level, and law enforcement, rather than to community-focused preparedness and response activities. For example, the President's 2003 budget allocates approximately 8% of homeland security funding for training local "first responders" – fire, police, and sheriff's department personnel – as compared with 28% for border security.

Plans also include very modest amounts of funding for a "Citizen Corps" made up of local volunteers. In contrast, "E-government" and other IT-related initiatives are slated to receive substantial amounts of funding. The Department of Defense is the single largest recipient of funds in the proposed homeland security budget, followed by the Departments of Transportation and Justice.

As these budget figures suggest, emerging homeland security threats provide major revenue opportunities both for the ailing information technology industry and for defense, intelligence, and national security agencies and contractors. The notion that the control of future threats will be achieved through successful application of information technologies is accepted uncritically, in part due to the myth that technology solves all problems, and in part as a consequence of the vigorous marketing efforts on the part of the IT industry that began immediately after the tragic events of September 11. Commenting in February, 2002 on the events of September 11 as a bonanza for the industry, one technology newsletter observed that

> The Sept. 11 terrorist attacks have spurred the biggest push yet toward a vision of seamless, electronically integrated government that tech firms have spent years pitching to federal agencies . . . In the four months since the attacks, dozens of technology companies have tried to sell themselves as the answer to the government's security prayers. Suddenly, projects like a multi-million dollar database integration effort for the entire federal law enforcement community, explosive detection equipment for luggage screening at every U.S. airport and new surveillance technologies for the domestic war on terrorism are all possible, and they're all business (Harris, 2002, p. 1).

Another newsletter reports that in the rush to identify IT solutions to the terrorist threat

> One of the first proposals out of the gate was from Oracle . . . Almost immediately after the September 11 attacks, Chief Executive Larry Ellison began calling in television interviews and newspaper op-ed pieces for a national identification-card system . . . Oracle is also teaming up with Sun Microsystems, Electronic Data Systems, and PwC Consulting to help the Transportation Security Agency and other federal agencies adopt biometric technology to assess security risks of airline passengers and airport employees (Gilbert, 2002, p. 2).

With these and other new developments, preparedness for emerging threats is being framed not in terms of local capacity-building or developing resilience within civil society institutions, but rather as a challenge best addressed by private industry and by institutions that operate on the basis of command-and-control and secrecy. Disaster myths, such as the notion that in the event of a future attack residents of affected areas will panic and present problems for emergency responders, resonate with the domain claims of these organizational actors.

Like the movement toward command and control, the increased emphasis on specialized credentials and qualifications in the field of emergency management also serves to sharpen the boundaries between those who are seen as qualified to play a role in disaster management and presumably unqualified (and panicky) members of the public. The last fifteen years have been marked by the emergence of emergency management as a profession requiring a distinct set of skills, as indicated by the proliferation of specialized course curricula and degree programs and by the development of credentialing bodies in the emergency management field. This process of professionalization has been judged by disaster scholars both as broadening participation in disaster management activities to include more involvement by civil-society institutions and as a counterforce against centralized and hierarchical approaches to the management of disasters. For example, Sutphen and Waugh (1998, p. 9) see professionalization as "encouraging the 'de-militarization' of the field, reorienting agencies from their earlier 'command and control' orientation to more coordinative and cooperative orientations." However, after 9–11, as federally-initiated homeland security initiatives increasingly take precedence and exert influence over local emergency management systems, and as emergency management becomes "remilitarized," conceptions of what constitute appropriate professional qualifications and activities for emergency managers are also likely to change.

As new expertise-related claims are pressed and as new institutional alignments form around top-down command-and-control initiatives, the roles that are envisioned for those outside systems of command and control and for non-experts – particularly community residents and local non-governmental organizations – shrink. Systems for managing emerging threats are increasingly becoming the domain of government agencies and other organizations that can claim expertise in the fields of military and national security intelligence, information technology, and law enforcement. In the aftermath of September 11 and as a consequence of evolving definitions of societal danger, new institutional actors are gaining increasing influence over the activities and missions of state and local emergency management agencies in ways that will ultimately affect how those agencies plan for and respond to crisis events of all types.

Recycled disaster myths both reflect and provide justification for this larger trend, signaling the transformation of U.S. emergency management into a system based on a hierarchical military and law enforcement model, rather than on a more open and collaborative one. The resurgence of discredited views on disaster management such as the panic myth accompanies a broader shift in the perceived disaster roles of ordinary community residents and civil society institutions. Especially with respect to emerging threats, rather than being characterized as a resource in disaster situations (which is what both experience and social science research have shown them to be), members of the public are viewed increasingly as sources of disruption and as lacking the kinds of expertise needed to play an active part in disaster response. Crisis preparedness and response are now being redefined as the domain of agencies and organizations that have a command over technology and that can function well in an atmosphere requiring high levels of secrecy. Indeed, the presence of non-experts and non-credentialed responders at any disaster site may now well be recast as a security risk. Seen in this light, recycled myths serve a dual purpose: they provide a basis for placing the management of new dangers more solidly in the hands of institutions and professions can claim special expertise – particularly expertise with respect to technologies that support command and control – and at the same time, they provide a justification for not including the public more directly in planning for future disasters.

NOTES

1. Often these initial response efforts are carried out under extremely dangerous conditions. For example, in the earthquake that struck the San Francisco Bay Area on October 17, 1989, workers from a nearby facility climbed onto the collapsed Nimitz Freeway structure in West Oakland to help bring trapped motorists to safety, ignoring the danger posed by potential aftershocks that could have caused additional collapses (see sources cited in Tierney, 1994).

2. Arguments for expanding the role of the military were outlined in the official assessment of federal emergency management capabilities that was carried out by the National Academy of Public Administration following Hurricane Andrew. That report took a position against assigning a greater role for the military in domestic disaster response (National Academy of Public Administration, 1993).

3. Data on these and other activities that were carried out following the Trade Center attacks were collected through quick-response observational field work and the analysis of documents and media accounts undertaken by the staff of the Disaster Research Center. More detailed information can be found in Kendra and Wachtendorf (2001a, b, 2002).

4. The GDIN task force was co-chaired by representatives from the Central Intelligence Agency and the National Oceanic and Atmospheric Administration. The task force membership included personnel from the Department of Defense, the National Security Agency, the National Imagery and Mapping Agency, the Defense Information Systems Agency, the National Aeronautics and Space Administration, and the Central Intelligence Agency, along with representatives from the Federal Emergency Management Agency, the Environmental Protection Agency, and other science-oriented agencies such as the National Science Foundation.

ACKNOWLEDGMENTS

The World Trade Center crisis response research discussed in this paper is supported by grants from the National Science Foundation, supplement to grant no. EEC-9701471, the Multidisciplinary Center for Earthquake Engineering Research, an NSF-sponsored engineering education and research center headquartered at the State University of New York at Buffalo, and the Public Entity Risk Institute. I wish to thank Tricia Wachtendorf, Disaster Research Center field director, and James Kendra, DRC's research coordinator, for the data and information they provided for this paper. Thanks are also due to Charles Perrow and Lee Clarke, who offered helpful critiques on an earlier version of the paper. The views expressed here are mine alone.

REFERENCES

Aguirre, B. E., Wenger, D. E., & Vigo, G. (1998). A test of the emergent norm theory of collective behavior. *Sociological Forum, 13*, 301–320.

Aguirre, B. E., Wenger, D., Glass, T. A., Diaz Murillo, M., & Vigo, G. (1995). The social organization of search and rescue: Evidence from the Guadalajara gasoline explosion. *International Journal of Mass Emergencies and Disasters, 13*, 93–106.

Atsumi, T., Sugiman, T., Mori, H., & Yatsuduka, I. (1996). Participant observations on volunteer organizations emerging after the Great Hanshin earthquake: Case of the Nishinomiya volunteer network and the local NGOs coordinating team for the Great Hanshin earthquake. In: *Proceedings of the International Conference on Water Resources and Environmental Research* (Vol. 2., pp. 455–462).

Beamish, T. D. (2002). Waiting for crisis: Regulatory inaction and ineptitude in the Guadalupe Dunes oil spill. *Social Problems, 49*, 150–177.

Biohazard News (2001). Interview with Thomas A. Glass: Public health response to mass casualty disaster. December 5, 1–6.

Cauchon, D. (2001). For many on September 11, survival was no accident. *USA Today*, December 19, 1A, 3–4A.

Centers for Disease Control and Prevention (1999). Bioterrorism Readiness Plan. Available on the Web at http://www.cdc.gov

Clarke, L. B. (1989). *Acceptable risk? Making decisions in a toxic environment.* Berkeley: University of California Press.

Clarke, L. B. (2002). Panic: Myth or reality? *Contexts* (Fall), 21–26.

CNBC, Inc. (2002). Transcript from "Hardball with Chris Matthews," March 4.

Comfort, L. (1999). *Shared risk: Complex systems in seismic response*. Oxford, UK: Pergamon. Elsevier Science Ltd.

Disaster Information Task Force (1997). *Harnessing information technology for disaster management.* Washington, DC: Global Disaster Information Network Task Force.

Drabek, T. E., & Haas, J. E. (1969). Laboratory simulation of organizational stress. *American Sociological Review, 34*, 223–238.

Dynes, R. R. (1993). Disaster reduction: The importance of adequate assumptions about social organization. *Sociological Spectrum, 13*, 175–192.

Dynes, R. R. (1994). Community emergency planning: False assumptions and inappropriate analogies. *International Journal of Mass Emergencies and Disasters, 12*, 141–158.

Dynes, R. R., Quarantelli, E. L., & Wenger, D. (1990). *Individual and organizational response to the 1985 earthquake in Mexico City*. Mexico. Newark, DE: University of Delaware, Disaster Research Center.

Fischer, H. W., III (1998). *Response to disaster: Fact vs. fiction and its perpetuation: The sociology of disaster*. New York: University Press of America.

Gellman, B. (2002). Government is slow to offer safety plans. Local, National offices have yet to disclose advice people could use in a terrorist attack. Washington Post Staff Writer, Tuesday, August 6, p. A01.

Gilbert, A. (2002). Tech companies chase homeland security. News.com. July 12.

Glass, R. A., & Schoch-Spana, M. (2002). Bioterrorism and the people: How to vaccinate a city against panic. *Clinical Infectious Diseases*, *34*, 217–223.

Harris, S. (2002). Tech insider: The homeland security market boom. GovExec.com. February 11, 2002.

International Federation of Red Cross and Red Crescent Societies (1996). *World disasters report 1996*. New York: Oxford University Press.

Johnson, N. R. (1987). Panic at the 'Who concert stampede': An empirical assessment. *Social Problems*, *34*, 362–373.

Johnson, N. R., Feinberg, W. E., & Johnston, D. J. (1994). Microstructure and panic: The impact of social bonds on individual action in collective flight from the Beverly Hills Supper Club fire. In: R. R. Dynes & K. J. Tierney (Eds), *Disasters, Collective Behavior, and Social Organization* (pp. 168–189). Newark, DE: University of Delaware Press.

Johnston, D. M., & Johnson, N. R. (1989). Role expansion in disaster: An investigation of employee behavior in a nightclub fire. *Sociological Focus*, *22*, 39–51.

Kendra, J., & Wachtendorf, T. (2001a). *Elements of community resilience in the World Trade Center attack*. Newark, DE: Disaster Research Center, University of Delaware. Preliminary Paper No. 318.

Kendra, J., & Wachtendorf, T. (2001b). *Rebel food ... renegade supplies: Convergence after the World Trade Center attack*. Newark, DE: Disaster Research Center, University of Delaware. Preliminary Paper No. 316. Paper presented at the annual meeting of the American Sociological Association, Chicago, IL, August 16–19.

Kendra, J., & Wachtendorf, T. (2002). Creativity in emergency response after the World Trade Center attack. In: *Proceedings of the 9th Annual Conference of the International Emergency Management Society* (pp. 69–81). Waterloo, Canada, May 14–17.

Kreps, G. A. (1990). The federal emergency management system in the U.S.: Past and present. *International Journal of Mass Emergencies and Disasters*, *8*, 275–300.

Langewiesche, W. (2002). *American ground: Unbuilding the World Trade Center*. New York: North Point Press.

McKinsey and Company (2002a). Improving NYPD emergency preparedness and response. *Report to the New York City Police Department*. New York: McKinsey and Company.

McKinsey and Company (2002b). *Increasing FDNY's preparedness*. New York: McKinsey and Company.

Mitchell, J. T., Thomas, D. S. K., Hill, A. A., & Cutter, S. L. (2000). Catastrophe in reel life vs. real life: Perpetuating disaster myth through Hollywood films. *International Journal of Mass Emergencies and Disasters*, *18*, 383–402.

National Academy of Public Administration (1993). *Coping with catastrophe: Building an emergency management system to meet people's needs in natural and manmade disasters*. Washington, DC: National Academy of Public Administration.

New York Times (2002a). *Report on 9/11 finds flaws in response of Police Dept.* July 27, A1.
New York Times (2002b). *Fire Dept. lapses on 9/11 are cited.* August 3, A1.
Noji, E. K. (1989). The 1988 earthquake in Soviet Armenia: Implications for earthquake preparedness. *Disasters, 13*, 255–262.
NYCStories.com (2001). Story posted by Justin Gauvin, describing his experience being transported by ferry from Midtown Manhattan to New Jersey on the day of the attack.
Podolny, J. M., & Page, K. L. (1998). Network forms of organization. *Annual Review of Sociology, 24*, 57–76.
Quarantelli, E. L. (1985). Realities and mythologies in disaster films. *Communications, 11*, 31–44.
Quarantelli, E. L., & Dynes, R. R. (1972). When disaster strikes (it isn't much like what you've heard and read about). *Psychology Today, 5*, 66–70.
Snyder, J. (2001). Ferries to the rescue after World Trade Center terror attack. Marinelog.com. No date.
Stallings, R., & Quarantelli, E. L. (1985). Emergent citizen groups and emergency management. *Public Administration Review, 45*, 93–100.
Stravelli, G. (2001). *Ferries deliver aid during, after tragedy at WTC: Transport victims then support rescue effort.* Atlanticville, September 20.
Sutphen, S., & Waugh, W. L., Jr. (1998). Organizational reform and technological innovation in emergency management. *International Journal of Mass Emergencies and Disasters, 16*, 7–12.
Symons, J. (2001). Report on Royal Society and British Association for the Advancement of Science meeting on bioterrorism. Report prepared for Internet newsletter, http://www.worldfoodscience.org
Tierney, K. J. (1994). Emergency preparedness and response. In: *Practical Lessons from the Loma Prieta Earthquake* (pp. 105–128). Washington, DC: National Academy Press.
United Nations Centre for Regional Development (1995). Comprehensive study of the great Hanshin earthquake. Nagoya, Japan: United Nations Centre for Regional Development, Research Report Series No. 12.
Washington Post (2001). U.S. details response to smallpox. November 27, A01.
Waugh, W. L., & Sylves, R. T. (2002). Organizing the war on terrorism. *Public Administration Review, 62*, 81–89.
Weinstein, I. (1947). An outbreak of smallpox in New York City. *American Journal of Public Health* (November).
Wenger, D. E., Dykes, J. D., Sebok, D., & Neff, J. L. (1975). It's a matter of myths: An empirical examination of individual insight into disaster response. *Mass Emergencies, 1*, 33–46.
Wenger, D. E., James, T. F., & Faupel, C. E. (1985). *Disaster beliefs and emergency planning.* New York: Irvington.
Wise, C. R. (2002). Organizing for homeland security. *Public Administration Review, 62*, 131–144.
Wood, D. (2002). America is dangerously vulnerable to panic in terror attack, experts say. http://www.newhousenews.com

THE FOX AND THE HEDGEHOG: MYOPIA ABOUT HOMELAND SECURITY IN U.S. POLICIES ON TERRORISM[1]

James K. Mitchell

ABSTRACT

Following the disasters of 9/11/01 the U.S. government has embarked on what is intended to be a comprehensive response to the hazard of further terrorist attacks on Americans at home and abroad. This paper addresses the homeland component of the response and asserts that both the general approach and the measures being deployed are neither comprehensive nor well-balanced. The broad goal of security is losing ground to the narrower objective of defense; mitigation strategies are being overshadowed by preparedness and response alternatives; expert systems are preferred over grass-roots bottom-up ones; and possibilities for reducing human vulnerability are being ignored in favor of programs that aim to reduce risks or lessen the vulnerability of built structures and infrastructures. Preferences for the use of sophisticated technologies that are intended to quarantine terrorism and minimize its consequences far outnumber efforts to engage with the messier realm of ideas and behaviors related to terrorism. Yet it is the latter that shape the public interpretation of terrorism risks,

Terrorism and Disaster: New Threats, New Ideas
Research in Social Problems and Public Policy, Volume 11, 53–72

ISSN: 0196-1152/doi:10.1016/S0196-1152(03)11005-8

structure patterns of exposure and affect the coping capabilities of threatened communities. Without substantial changes to policy that take account of these deficiencies, Americans are likely to find themselves little better prepared to confront the challenges of future terrorist attacks on targets in U.S. territory and the nation's ability to address other kinds of hazards may be seriously compromised.

CROSSING THE ARCHILOCHIAN DIVIDE

The fox knows many things, but the hedgehog knows one big thing.
Archilochus, Greek lyric poet (710–676 BCE).

No one can be sure what Archilochus meant when he wrote these lines but they have been interpreted by many, including the British philosopher Sir Isaiah Berlin (1951), to illustrate the contrast between those who favor pluralistic outlooks and those who are committed to a single worldview. Such a dichotomy provides an apt metaphor for the general transformation of the U.S. political agenda in the aftermath of 9/11, wherein a complex mix of existing national priorities has been displaced by a pre-eminent War on Terror. However, the metaphor also captures another kind of transformation that has been proceeding in a less publicized manner. This involves the crafting of national policies and programs that are intended to protect Americans in the homeland. Homeland Security policy makers have ignored interpretations of contemporary terrorism as a symptom of broad societal conflicts, such as those that are associated with globalization, and have chosen to view it as a narrower kind of human pathology. Instead of adding adjustments that might mitigate societal tensions (e.g. by modifying institutional structures, imagery and belief systems that encourage unnecessarily rancorous political, economic and cultural changes – both at home and abroad) they prefer to rely more or less completely on heightened civilian vigilance combined with a limited range of domestic surveillance, preparedness and emergency response measures.

This paper suggests that such a narrow approach is not in the best interests of the American public (nor in the interests of any population that is similarly threatened) because it does not address fundamental aspects of homeland terrorism and ignores potentially valuable contributions to homeland security. It may also have deleterious effects on programs for reducing other kinds of environmental hazards that fall under the jurisdiction of agencies involved in implementing the present antiterrorism strategy.

U.S. HAZARD MANAGEMENT INITIATIVES BEFORE 9/11

The United States has previously faced few major threats of catastrophic terrorism within the country's borders so the national hazards management system has mainly been designed to address natural and technological emergencies as well as epidemic diseases and other health crises. Out of this experience has come a large fund of knowledge about environmental risks and human responses as well as a set of distinctive public policies and institutional arrangements.

The system that has evolved is multifaceted and administratively fragmented, with responsibilities shared among separate federal agencies, different level of government, public-private partnerships, non-governmental entities and – perhaps most important – individuals, families, neighbors and friends. By the end of the 20th century, U.S. approaches to managing natural and technological hazards combined a broad range of means in pursuit of varied – but mutually supporting – ends. Agents of hazard were addressed as well as vulnerable populations; anticipatory policies and programs complemented reactive ones; efforts to engage the underlying causes of hazard were coupled with palliative responses; and behavioral strategies were pursued in consort with coping technologies. Decision-making was increasingly decentralized outwards from government to civil society and devolved downwards from the federal level to the local level, while "partnership" had become an ever-present watchword in most sectors of the hazards community. Variously successful efforts were also made to shift more of the burden for protecting against hazard from public agencies to the private sector via widespread criticism of publicly-sponsored disaster relief (Platt, 1999) and promotion of insurance as an alternative that might permit individuals to choose their own levels of (economic) protection. Though the system was by no means perfect it addressed the multiple faces of hazard and the needs of diverse stakeholders. It was also a major improvement on previous policies that had not attempted to engage the full spectrum of hazard issues or to employ the full range of useful adjustments (Mileti, 1999; National Research Council, Natural Disasters Roundtable, 2002).

Some examples of existing natural hazard initiatives, drawn from a plethora of available alternatives, include the following:

- Legislation that addresses hazards by sector (e.g. National Earthquake Hazard Reduction Act) and by function (e.g. Stafford Disaster Relief Act);
- Administrative programs for single hazards like floods, that use single means such as structural protection works (e.g. National Dam Safety Program) or for

multifaceted issues like hazard zone occupance, that use multiple means such as agency-based floodplain management systems or inter-agency task forces (e.g. U.S. Flood Plain Management Task Force);

- Research and information organizations like universities, national laboratories, think tanks, consulting firms and professional associations that study hazards or processes of risk assessment or systems of hazard management, and disseminate findings to specialized users as well as the lay public;
- Non-governmental bodies like the American Red Cross, or the Institute for Business and Home Safety, that organize and publicize knowledge about risks, train hazard professionals or volunteers, and seek to change behaviors that place people and investments at risk or keep them from adopting safer alternatives;
- Public-private partnerships that bring together scientists, investors, victims, advocates, regulators, public interest groups and other "stakeholders" to analyze and/or manage shared hazards (e.g. New York City Consortium for Earthquake Mitigation);
- Private voluntary relief organizations and government assistance agencies that act in coordination with FEMA to direct aid to disaster victims and impacted communities;
- Fire, police, emergency medicine and other public emergency services that plan for and respond to disasters or threatened disasters;
- Science-based agencies like the U.S. Geological Survey and the National Weather Service that study, monitor, predict and forecast geological, hydrological or atmospheric risks and – in cooperation with state, local and non-profit organizations – issue warnings to affected publics;
- Mass media news and entertainment organizations that transmit hazard warnings and help to shape public perceptions of risks and vulnerabilities in newspapers, magazine, television, radio, movies and the Internet;
- Technology-based agencies like the National Bureau of Standards or regional building code associations that seek to improve hazard-resistant construction practices;
- Insurance agencies like the Federal Insurance Administration, that – in partnership with local governments – oversees mapping, and regulation of floodplains and coastal high hazard areas via policies that are variously intended to increase awareness of flood risks, ease the recovery of flood victims and reduce long-term human settlement of areas exposed to significant flood risks.

Before 9/11 the Federal Emergency Management Agency (FEMA) played a leading role in the U.S. hazard management system. Its chief functions had become the coordination of federal responses to all kinds of major disasters and the promotion of improved public policies for reducing losses. FEMA was itself a product of at least thirty years of tortuous governmental efforts to find a satisfactory way of addressing problems of disaster. These efforts bore fruit when a narrow civil defense interpretation of the agency's mission was rejected in favor of a much broader strategy that combined reactive policies with anticipatory ones[2] (Platt, 1999; Popkin, 1990; Witt & Morgan, 2002).

FEMA was one of the success stories of federal governmental reform during the 1990s, an achievement that was repeatedly acknowledged by President Clinton in several State of the Union messages. This status was earned in a number of ways including by streamlining the delivery of disaster services, improving coordination with state and local governments and fostering broad partnerships with NGOs, private sector organizations and public bodies in disaster-affected communities. But its most important initiative was the promotion and adoption of hazard mitigation policies and programs (Godschalk et al., 1999). As noted in a 1995 report, "For the first time in the history of federal disaster assistance, mitigation – sustained action taken to reduce or eliminate long-term risks to people and their property from hazards and their effects, has become the cornerstone of emergency management" (FEMA, 1995). In other words, there was a shift away from reactive policies that treat symptoms towards anticipatory strategies that are designed to affect underlying causes of hazard.

RECENT CHANGES OF EMPHASIS IN PUBLIC POLICIES: FROM SECURITY AND MITIGATION TO DEFENSE AND EMERGENCY MANAGEMENT

Some FEMA leaders recognized that 9/11 carried the potential for revolutionizing management of hazards and hoped that it might usher in a new paradigm of sustainable hazards mitigation (Armstrong, 2002). Instead, only those aspects of mitigation that apply to the built environment are addressed in the emerging Homeland Security strategy (FEMA, 2003). The mitigation of human risks and vulnerabilities is not considered directly. In addition, FEMA has been placed in a new Department of Homeland Security together with units like the Secret Service, Coast Guard and Border Patrol that are primarily concerned with surveillance, emergency preparedness and response (http://www.whitehouse.gov/deptofhomeland/toc.html). James Lee Witt, FEMA's former director, advised against this move, and local emergency management officials – who generally are enthusiastic about broad mitigation and prevention programs – are reported to be concerned (GovExec.com, November 15, 2002). Despite assurances to the contrary, external observers might be forgiven if they once again construed FEMA as an Emergency Preparedness and Response agency. Moreover the bulk of that restricted focus seems likely to be directed towards just one kind of emergency – terrorist attacks. This will undercut FEMA programs that are intended to reduce the effects of other kinds of hazards.

The narrowing of perspectives that is detectable in the overshadowing of mitigation as a lead policy also shows up in the terminology that is used by stakeholders to characterize the Homeland initiative. Many such groups routinely use the label "homeland defense" to describe their interests and involvements rather than the broader-sounding "homeland security." Among others these include the U.S. Departments of State and Defense, the Center for Strategic and International Studies, The Heritage Foundation, The Brookings Institution, The National Academy of Engineering and publishers of the on-line Homeland Defense Journal. This limited purview stands in contrast to the broad interpretations of security that are prevalent among national and international organizations concerned with development, environment, refugees, human rights, disaster reduction, health care, social welfare and other issues (Harvard School of Public Health, 2003). The latter point up notions of human security rather than state security, look toward the provision of basic human needs – not the validation of national policy agendas, and give pride of place to a "culture of prevention" instead of a "culture of defense" (International Institute for Sustainable Development, 2003; Moller, 1993; Organization for Economic Cooperation and Development, 2000). The differences in emphasis are palpable and portentous. At a time when a new kind of hazard calls for creative and ambitious coping policies, they signal a loss of confidence in non-militarized problem solving capabilities. Without a prevention culture, it is a moot question whether the existing Homeland strategy will be sufficient to quell the anxieties of Americans and keep losses within acceptable bounds.

A blinkered perspective on security against terrorism is also evident in the governmental reorganization that created the Department of Homeland Security. Some of the public agencies that share responsibilities for natural, technological and health emergencies have been incorporated into the new department or are closely involved in antiterrorism planning and response efforts (e.g. Federal Emergency Management Agency, U.S. Coast Guard, Centers for Disease Control and Prevention). But others, that have made equally important contributions to improved management of catastrophic environmental risks, remain outside; outside not just the Department of Homeland Security but also on the outer fringes of the larger Homeland Security Strategy. (For example, U.S. Geological Survey, National Oceanic and Atmospheric Administration-National Weather Service; U.S. Army Corps of Engineers; most of the National Laboratories.) In other words, the government is not making the most of its own valuable hazard management expertise.

Moreover the homeland security initiative has reached out for guidance from the hazard research community in a quite selective manner. Expertise from certain of the physical sciences, medicine, engineering and social sciences has been solicited (e.g. National Research Council, Committee on Science and Technology for Countering Terrorism, 2002), yet inputs from several vitally important fields are either missing or muted. Among others these include – risk assessment, disaster research, hazards research and sustainability studies (Anderson, 2002; Chess & Calia, 2002; Cutter, Richardson & Wilbanks, 2003; Desiler, 2002; Haas, 2002; Kunreuther 2003a, b; Lave, 2002; Pate-Cornell, 2002; Schwing, 2003; Slovic, 2002; Wisner, 2002). Programs that address hazards in the context of issues like climate change, sustainability and urban governance (Heiken et al., 2000; Intergovernmental Panel on Climate Change, 2001; Kasperson & Kasperson, 2001a) could make valuable contributions to the implementation of terrorism vulnerability-reduction measures but they – and their like – are rarely to be found in the prevailing discourse on homeland security. Obviously, the Homeland Security Strategy is still evolving but these initial signs raise concerns about what is to come. Compared with the conceptual breadth and diversity of programs for the reduction of natural and technological hazards, Homeland Security policy exhibits a narrowing of vision and a preference for measures whose scope of application and record of effectiveness are both quite limited.

In view of the preceding signals, national policy on hazards management seems to be reverting to an earlier (Cold War) era. This does not mean that the concerns of that era – surveillance, emergency preparedness and response – or protective measures associated with them, lack value as adjustments to terrorism. They surely remain indispensable for coping with unexpected fast-developing hazards, but they are also incomplete and temporary answers to complex problems like terrorism that continue to fester if underlying causes are left unattended. Public reaffirmation of the need to construct public policy on a firm base of knowledge about hazards would be an appropriate initial step toward correcting this imbalance.

INTENDED ACTIONS: HOMELAND SECURITY BUDGETS[3] AND PROGRAMS

In two successive years the President of the United States has requested \$37.7 Billion and \$41 Billion respectively for support of efforts to carry out the domestic part of the nation's War on Terrorism (Fig. 1). These huge sums dwarf the combined budgets of agencies that have traditionally been responsible for addressing natural and technological disasters. Most of the money is allocated for developing and staffing new or improved information and technology systems that are intended

Spending category	US$ Billions (109)
"First responder" needs	3.5
Bio-terrorism initiatives	5.9
Security of U.S. borders	11.0
21st century technology	17.3
Total	37.7

Fig. 1. U.S. Homeland Security Budget FY 2003. *Source:* http://www.whitehouse.gov/homeland/homeland_security_book.html

to detect potential terrorists, fight the health effects of bio-terrorism, ensure the safety of critical infrastructures, and help emergency managers respond to attacks that actually occur. A much smaller but difficult-to-measure sum is earmarked for reducing the vulnerability of so-called "critical infrastructures" (e.g. water supply networks, energy production & supply lines, data and information systems).

Sixty-nine percent of budget funds are intended for agencies that guard U.S. land and water borders or transportation systems. Emergency Preparedness and Response consumes a further 17%. Less than 5% is allocated for information analysis, infrastructure protection, science and technology, with the balance to miscellaneous other – mainly housekeeping – purposes (*New York Times*, February 4, 2003). However, much of the funding for emergency management will actually consist of training, equipment and pass-along programs that target the two million-strong body of emergency "first responders" (i.e. fire, police, medical and rescue services).

Technological fixes are strongly preferred. Among these are the following: personal protective equipment; chemical and biological detection systems; interoperable communications gear; a National Pharmaceutical Stockpile; new vaccines, medicines and diagnostic tests; technologically advanced vehicle and ship tracking equipment; an Entry-Exit Tracking System for foreign visitors; enhanced defenses for critical high-risk vessels and coastal facilities; 21 e-government initiatives that will provide better protection of databases; a National Infrastructure Protection Center; a Cyberspace Warning Intelligence Network, a wireless priority access program for certain cellphone users; a National Infrastructure Simulation and Analysis Center and a national "Cybercorps." The scale of the projected spending is such that it presents defense industry corporations with greatly expanded opportunities to secure lucrative contracts for research, development, production, installation and operation of new technology systems (*New York Times*, March 20, 2002, p. C1, C12).

Present budget spending categories are strongly driven by a concern for identifying and managing risks associated with terrorism and for coping with threatened or actual attacks. Policy makers hope to keep terrorists and their weaponry at arms length, preferably outside U.S. borders where they might be dealt with by military means but – failing that – by superior intelligence and detection capabilities within the homeland. If these measures prove inadequate, emergency workers and emergency management systems are expected to minimize subsequent losses. This can be described as a top-down, expert-led, directive approach for the implementation of risk-reduction measures backed up by improved post-attack responses. But one of the central lessons of the American experience of coping with natural and technological hazards is that bottom-up, lay-inspired, voluntaristic approaches have been just as important, especially when applied to the reduction of vulnerability.

THE IMPORTANCE OF VULNERABILITY[4]

$$\text{Hazard} = (\text{Risk}) \times (\text{Vulnerability})$$

$$\text{Vulnerability} = (\text{Exposure}) \times (\text{Resistance}) \times (\text{Resilience})$$

It is well known that, in every population affected by hazards, some people are better able to absorb losses and bounce back from them than are others. This capacity has been labeled "vulnerability" although the meaning of that term to hazard specialists goes beyond vernacular connotations. "Vulnerable" is derived from the Latin for "wound" and can roughly be translated as the potential for experiencing loss. As used in connection with natural hazards, "vulnerability" includes the interrelated notions of exposure[5] resistance[6] and resilience.[7] In other words, it indicates the human capacity for responding to loss as well as the potential for experiencing loss. Populations that combine high exposure to risk, low levels of resistance and weak resilience are most vulnerable to hazards.

Although the concept of vulnerability is most commonly associated with constraints on the agency of social groups (Blaikie, 1994; Enarson & Morrow, 1998; Fothergill, Maestas & Darlington, 1999), it also applies to individuals, to non-human entities (e.g. plants, animals, buildings) and to social structures or environmental contexts. For example, institutions like families, voluntary organizations, communities, markets, firms, public agencies and governments are differentially vulnerable. So too are all kinds of human and technological systems, including those that make possible the operation of complex urban communities (e.g. buildings, data banks, utility networks). When many different groups are engaged in the simultaneous or sequential use of the same location, the vulnerabilities of place and time may be more useful analytical constructs than those of people and objects (Cutter et al., 2001; Hewitt, 1997; Kasperson & Kasperson, 2001b).

Vulnerability may sometimes be a function of individual behavior but it is just as often a byproduct of societal processes that have differential effects on human populations; processes like – knowledge production, information flows and communication networks; patterns of migration; demographic trends; the structure and performance of economic systems; status, power and influence relationships; access to opportunities for advancement; belief systems, etc. Whereas risk-reduction strategies usually seek to modify the phenomena that inflict losses (e.g. storms, fires, explosions) or to separate them from populations at risk, vulnerability-reduction strategies focus on improving the capabilities of affected populations to avoid, prevent, resist, absorb and recover from loss. (In a general sense the difference is between keeping threats away from people – risk reduction, vs. keeping people away from threats – vulnerability reduction.)

Vulnerability may be reduced in many ways. Most of those that have been adopted by government agencies involve training, prediction, evacuation and structural strengthening.[8] But the biggest opportunities for reducing vulnerability (and for bringing down losses generally) focus on changing knowledge, behavior and societal impediments to protective action. Some of the ways the latter may be accomplished include:

- Improving information upon which decisions are based by:
  - Undertaking research that identifies and analyzes vulnerabilities and vulnerability-producing mechanisms;
  - Raising awareness of vulnerabilities among policy makers and the general public;
  - Altering attitudes or beliefs that affect vulnerability and security;
  - Providing exposed populations with accurate information about vulnerabilities and appropriate adjustments.
- Promoting a wider range of choices that reduce vulnerability by:
  - Developing innovative institutions and technologies;
  - Facilitating collective action by separate stakeholders;
  - Modifying societal structures that fail to take account of vulnerability or that increase vulnerability;
  - Removing barriers to the adoption of appropriate prevention, avoidance or protection measures;

- ◦ Ensuring that the most vulnerable have the resources to adopt needed adjustments.

Though risk-reduction has tended to have first claim on the attention of policy makers and hazards managers in the USA, vulnerability-reduction eventually becomes an essential complementary goal. The order in which these strategies are typically adopted probably owes something to the ease with which risk can be represented (in Western cultures) as an externalized "other" that threatens a neutral "us" and the slower-to-emerge, uncomfortable realization that humans themselves (including perpetrators, victims, helpers, leaders, and publics unaffected by hazard) construct the conditions that make for vulnerability. Policies that address only one set of variables (risks or vulnerabilities) are, at best, seriously compromised and, at worst, recipes for failure.

During the past decade vulnerability analyses have begun to show up as complements to risk analyses in hazard assessment procedures around the world. Organizations like FEMA, NOAA (National Oceanic and Atmospheric Administration), EPA (Environmental Protection Agency) and the IPCC (Intergovernmental Panel on Climate Change) have all developed their own variations and others are in the pipeline. For example, the UNDP (United National Development Program) is about to publish the first in an annual series of World Vulnerability reports that will provide guidance about vulnerability reduction in a broad array of economic development contexts. These initiatives variously incorporate, into the calculus of hazard, the human capacity to create or modify risks by consciously or inadvertently manipulating factors of exposure, resistance and resilience (IPCC, 1997). Initially, the interpretations of vulnerability that were employed were limited in scope, referring mostly to vulnerable structures like homes, business and utility lines – phenomena that could be mapped as fixed entities in a risk-prone environment. Gradually these minimalist interpretations are being augmented by broader analyses that measure the vulnerability of people – both at rest and in motion – first as individuals, then as genders, families, groups and institutions, as well as the vulnerability of other species and nonliving entities that are deemed of special value. Different kinds of functional vulnerabilities of human communities are now being uncovered (Mitchell, 2003b). Feedback processes that take account of synergisms among and between elements of society and nature are also being added to the mix.

GROWING NATIONAL AWARENESS OF "VULNERABILITY"

Since 9/11 it has been common to hear about a heightened sense of vulnerability that is experienced by Americans as well as a real shift by the country towards a more precarious and uncertain future. These sentiments are shared by people as different as the *New York Times*' columnist Frank Rich and the President of the United States:

> We live in a different America today than we did only the day before Tuesday. . . . we have lost our untroubled freedom of movement that we consider a birthright. We have lost our illusion of impregnability. But beneath those visceral imperatives an entire culture has been transformed. This week's nightmare, it's now clear, has awakened us from a frivolous if not decadent decade long dream, even as it dumps us into an uncertain future we had never bargained for.
>
> Frank Rich, 2001. *New York Times*, September 15.

> . . . the characteristics of American society that we cherish – our freedom, our openness, our great cities and towering skyscrapers, our modern transportation systems – make us vulnerable to terrorism of catastrophic proportions. America's vulnerability to terrorism will persist long after we bring justice to those responsible for the events of September 11.
>
> President's FY 2003, budget message.

When used in this way the underlying meaning of vulnerability seems to be a hidden state of passive fragility that has just recently been rendered visible by unforeseen events. This is far-removed from specialist understanding of the term. Neither is there much evidence that the framers of Homeland Security policies and programs recognize the importance or the complexity of the concept as so understood.

THE CONSTRUCTION OF VULNERABILITY TO TERRORISM

Vulnerability to terrorism is a direct outgrowth of how humans construct their societal institutions and how they conduct their daily lives. It reflects the choices that we as individuals make about where, when and how to reside, work, travel and relax as well as the local and global structures that society has evolved to accommodate those choices. Terrorists exploit the opportunities for inflicting harm that such choices create. Consider the example of commuting – a daily activity cycle that engages many millions of people.

Commuting involves a diurnal shift of population because jobs and residences are located in different parts of urban areas and workers must move between them. The forces that create these disparities of land use and activity have causal chains that can reach far across time and space and are deeply embedded in broad social, economic, political and cultural processes. The resulting movement of commuters brings people together in concentrated groups that have been favorite terrorist targets. Here the history of bombings and gas attacks on trains and stations in London, Tokyo, Paris and Bologna is dismally illustrative. At the urban end of the commute, individuals redistribute themselves among high-rise offices, public buildings, street level stores, restaurants, parks, entertainment complexes and tourist venues. These are the kinds of places favored by car bombers and suicide bombers, as residents of Bogota, Colombo, Jerusalem, Manchester, Moscow or Oklahoma City can clearly attest. (In September 2001 they were also the targets aimed at by airplane hijackers who struck New York and Washington, DC.) Though the United States has little experience of terrorist attacks in the residential neighborhoods to which commuters return in the evening, other countries do. Deliberately planted explosives have toppled apartment towers in Moscow and community centers in Buenos Aires; social clubs have been fire-bombed and machine-gunned in Belfast; pubs have been blown up in Birmingham, targeted victims and passersby engaged in leisure pursuits have died in shoot-outs, kidnappings and assassinations in Tijuana, Rome, Madrid, Athens and Cairo.

For individual commuters, types and levels of vulnerability to terrorism change from place to place, throughout the day and over longer periods. They also vary in accordance with the amount of knowledge about risks and vulnerabilities that commuters possess; the degree to which they can avail themselves of protective resources; their access to alternatives; and the willingness of society to facilitate different choices about where and how they work, live or recreate. Above the level of individuals, various kinds of collective vulnerabilities exist – each responsive to its own array of exposure, resistance and resilience factors.

Beyond the realm of dangers to life and property lie other – more systemic – vulnerabilities. For example, the cities that terrorists so often target are not just homes, workplaces and economic engines. They are also highly specialized ecosystems, learning environments, symbolic territories, performance spaces and muses. These and other urban functions are differentially affected by terrorism in ways that most people recognize as intuitively important but they are rarely taken account of by formal hazard-coping systems. Symbolic losses are an obvious case; it may matter little to the efficiency of a community that a historical monument or a community symbol is destroyed by terrorism but the psychic impacts of such a loss may be sufficient to galvanize reactive behaviors that greatly amplify the effect of the initial attack. Likewise, fear of further terrorism may alter the public "performances" that characterize urban street life or human interchanges of other sorts and are just as essential to the continuance of a city as reliable public services and safe buildings.

STUNTED CONCEPTIONS OF VULNERABILITY IN THE HOMELAND SECURITY PROBLEMATIQUE

In the present Homeland Security strategy, what little attention is devoted to vulnerability concerns the vulnerability of things rather than the vulnerability of people. Mostly the goal is to reduce the vulnerability of "critical infrastructures" such as pipelines, bridges, airports, electricity production and distribution systems as well as the electronic data networks that undergird telecommunications and the financial services industry[9] (Haimes & Longstaff, 2002). The intent is to minimize physical damage to public property and "loss of service" to users. Unfortunately, this leaves many aspects of vulnerability unaddressed.

Budget priorities pay very little attention to the vulnerabilities of Americans in their homes, neighborhoods, workplaces and leisure surroundings or to the potential for successfully adjusting private behaviors and public policies in light of new terrorism threats and the full range of appropriate responses.[10] Nor do they take account of the fact that, like the global society as a whole, vulnerabilities in different parts of the world are increasingly interconnected.

To a significant degree the President's budget priorities reflect decisions to treat terrorism in the homeland as if it were an infectious disease whose periodic outbreaks must be identified, isolated, controlled and preferably eliminated. The experience of anthrax attacks during 2001 and the advice of experts on biological warfare and public health (e.g. Centers for Disease Control and Prevention) were undoubtedly suggestive in this context. Such a model complements the military "War on Terror" that is being waged abroad and it is useful for understanding certain types of terrorism that might occur in the United States, especially those that involve biological weapons or the contagion-like spread of misinformation. But when applied more broadly, the limitations of the disease model are clear. Rather than resembling a biological process, terrorism is innately human in its causes and consequences. Its vulnerabilities (and risks) are human constructed; its aims and purposes are bound up with emotions, symbolism, ideas and values. Its targets are chosen as much for their psychological connotations as for their material ones. Reflexivity and social learning are key processes for perpetrators and victims alike. In other words, Homeland Security strategists need to take account of the expansive dimensions of socially constructed vulnerabilities, As part of that task it is time to consider how untapped resources might be mobilized both to reconceptualize the hazard of terrorism in ways that allow humans to live with increased uncertainties and to reduce the underlying vulnerabilities that contribute to a broad spectrum of hazard-related societal losses.

TOWARDS A VULNERABILITY-REDUCTION COMPONENT OF HOMELAND SECURITY STRATEGY

More can be done to reduce the vulnerability of people in the American homeland than is currently being undertaken. The details of an improved strategy will take time to work out but the following points might be useful points of departure.[11]

A first step is to recognize the complexity of vulnerability to terrorism and to plan accordingly in light of this knowledge.

Some of the complexity exists because factors that contribute to terrorism vulnerability in the homeland are often well separated from their effects in time, in space and in societal contexts. This calls into question a policy of strong dependence on domestic (i.e. within country – and especially local) capacities to analyze and address vulnerability issues.

Other aspects of complexity involve the setting of vulnerability thresholds and the relativity of multiple vulnerabilities. Vulnerability is not simply an absence of physical safety either of people or property. It is a relative potential for losses from different risks above a socially acceptable minimum. Humans are at risk to a wide suite of environmental threats, from natural extremes to accidents to diseases and beyond; some of these are included on official national agendas for action and others are not. Like risks, we choose the vulnerabilities and the levels of vulnerability that will trigger action as well as the ones that will be denied or subject to tacit indifference (Douglas, 1992; Douglas & Wildavsky, 1982). Clearly terrorism is now part of the U.S. national agenda but the levels of acceptable loss and the length and makeup of the total agenda are not at all obvious nor are there signs that a public dialogue on what the agenda should include is taking place. What other vulnerabilities might merit attention? Based on which indicators of expediency or severity? Using what criteria to take account of the differences between known chronic hazards and surprising acute ones? Should the others receive more or less attention than terrorism-related vulnerabilities?[12] Simultaneously or sequentially? As part of a narrow campaign against seemingly unambiguous agents that threaten immediate human safety at home or as a broader strategy against ambiguous structures and agents that undermine long and short-term bases of human security at home and abroad? Without a clear sense of what else merits societal response, not only does a war on the risks of terror become detached from a war on terrorism vulnerability but also from the context of other (comparative) hazards vulnerabilities. An inherently complex decision space would thereby be reduced to a menu of misleadingly simple choices among alternative risk-reduction technologies.

It is important that lay populations become active partners with governments and experts in efforts to reduce vulnerabilities (and potential losses) from terrorism.

Earlier it was pointed out, with respect to commuting, that the form, content and operational procedures of transportation modes, work spaces, living quarters and entertainment facilities used by commuters are products of decisions taken by diverse people who design, build, manage or occupy them. Therefore, it is reasonable to expect that efforts to reduce vulnerability will also need to involve many of the same groups. Not only will this enact the valuable principle of keeping responsibility for action in the hands of those who are likely to be most affected by terrorism, it will also help to dispel public illusions about the hazard that are based on fear of unknowns, it will temper discussions of security policy that might otherwise be ceded to unrepresentative activist groups with doubtful agendas, and it will engage a broad cross-section of society in shared projects that have therapeutic as well as protective value. Such a strategy requires a far greater level of shared knowledge about terrorism threats and vulnerabilities – and an openness to voluntary lay inputs – than presently exists. It also requires a willingness to take account of both the domestic aspects of vulnerability and those that are connected with the reach of U.S. institutions and influences outside the country.

CONCLUSION

In seeking to craft an effective Homeland strategy there can be no getting around the fact that security, mitigation and vulnerability are major – but poorly understood – components of the problem. Unfortunately they do not seem to have received the breadth of scrutiny and debate from researchers, policymakers and the general public that they deserve. Instead the emerging strategy is based on notions of terrorism that hark back to earlier and less productive ways of thinking about hazards.

The evolving policy for Homeland Security seems firmly rooted on the narrow side of the Archilochian dichotomy. More concerned about defense than security, tentative towards mitigation as a policy alternative, blinkered both about the concept of vulnerability and the use of vulnerability-reduction measures, it is also overwhelmingly reactive, palliative and reliant on technological fixes; in sum, deeply reminiscent of previous flawed (and sometimes quixotic) attempts to permanently eliminate specific hazards from the roster of public policy problems without addressing their causes. The federal government is now betting the collective security of Americans on stunted concepts and a truncated range of adjustments to hazard. Hard-won lessons about the value of broad human-centered approaches to hazards-reduction appear to have been forgotten or set aside in favor of narrowly targeted programs that objectify risks and place responsibility for public safety in the hands of technological elites or emergency experts. The hedgehog has, in effect, replaced the fox at the cutting edge of American hazard management. Such a change bodes ill for efforts to tackle the increasingly complex hazards of the 21st century, including terrorism. A reassessment of

Homeland Security policy, that makes room for foxes as well as hedgehogs, is urgently needed.

NOTES

1. Based on the author's presentation, "*Backing into vulnerability: Homeland security in the wake of 9/11*," Annual Meeting of the Association of American Geographers, New Orleans, March 5–9, 2003.

2. The changing names of FEMA's predecessors from 1953 to 1979 are indicative of this struggle: Federal Civil Defense Administration; Office of Civil Defense Mobilization; Office of Emergency Planning; Office of Emergency Preparedness; Defense Civil Preparedness Agency; Federal Disaster Assistance Administration; Federal Preparedness Agency.

3. At the time of writing details of the FY 2004 budget were not available; this discussion focuses mainly on the FY 2003 budget.

4. Some of this section is based on the author's contribution to the forthcoming *World Vulnerability Report*, United Nations Development Programme, Geneva.

5. Exposure is a measure of the population at risk.

6. Resistance is equated with the effectiveness of existing adjustments (i.e. measures that are intended to prevent, avoid or reduce losses). Non-functioning adjustments, or adjustments that fail, are indicators of high vulnerability.

7. Resilience refers to the capacity of a hazard-impacted entity to resume functioning in an acceptable manner after experiencing loss. Slow recovery times are usually indications of high vulnerability.

8. Properly functioning warning and evacuation systems can ensure that potential victims leave areas at risk and take shelter in buildings that are designed and constructed to enable them to withstand subsequent hazard shocks. The resilience of utility and transportation facilities or networks can be improved by planned rerouting and rescheduling of loads, installation of back-up devices, introduction of hazard-resistant design features and other modifications. Better training and mutual assistance arrangements among local emergency services can help to improve the volume and speed of response to disasters.

9. This approach is consistent with several historic precedents for federal involvement in the protection of infrastructure systems that were deemed essential to interstate trade and commerce (e.g. dredging and navigation improvements, canals, interstate highways).

10. Nor are they addressed by the growing number of private firms that offer security audits, threat assessments and other vulnerability-related services to business customers http://www.jbfa.com/terrorist-risk-vulnerability-assessment.html. When it comes to reducing the very real vulnerability of everyday behaviors it would seem that the general public is mostly being left to fend for itself.

11. Further information on implications for the hazards research community is contained in Mitchell (2002, 2003a).

12. As pointed out in a recent authoritative survey of 38,000 global respondents in 44 countries people in only a very few countries perceive terrorism as the leading national policy priority. For the poorest countries it is far outranked by issues like the availability of fresh water (Pew Research Center, 2002).

REFERENCES

Anderson, E. L. (2002). Assessing the risks of terrorism: A special collection of articles by former presidents of the Society for Risk Analysis. *Risk Analysis, 22*(3), 401–402.

Armstrong, M. J. (2002). *Urban hazard-mitigation: Increasing human security through re-assessing the steps toward livable communities*. Fairfax, VA: ICF Consulting. http://www.icfconsulting.com/Publications/doc_files/Urban_Hazard_Mitigation.pdf

Berlin, S. I. (1951). Leo Tolstoy's Historical Scepticism. *Oxford Slavonic Papers*, *2*, 17–54; reprinted with additions as The Hedgehog and the Fox: An Essay on Tolstoy's View of History (London, 1953: Weidenfeld & Nicolson).

Blaikie, P. et al. (1994). *At risk: Natural hazards, people's vulnerability, and disasters*. London: Routledge.

Chess, C., & Calia, J. (2002). Letter to the editor. *Risk Analysis*, *22*(6), 1039–1040.

Cutter, S. L. et al. (2001). *American hazardscapes: The regionalization of hazards and disasters*. Washington, DC: Joseph Henry Press.

Cutter, S. L., Richardson, D., & Wilbanks, T. J. (Eds) (2003). *The geographical dimensions of terrorism*. New York: Routledge.

Desiler, P. F., Jr. (2002). A perspective: Risk analysis as a tool for reducing the risks of terrorism. *Risk Analysis*, *22*(3), 405–413.

Douglas, M. (1992). *Risk and blame: Essays in cultural theory*. London: Routledge.

Douglas, M., & Wildavsky, A. (1982). *Risk and culture: An essay on the selection of technological and environmental dangers*. Berkeley: University Press of California Press.

Enarson, E., & Morrow, B. H. (Eds) (1998). *The gendered terrain of disaster: Through women's eyes*. Westport, CT: Praeger.

Federal Emergency Management Agency (1995). *Report of FEMAs Mitigation Directorate for Fiscal Year 1995*. Washington, DC.

Federal Emergency Management Agency (2003). *Hazard mitigation: Its not just for natural disasters anymore*. Washington, DC. http://www.fema.gov/fima/antiterrorism/

Fothergill, A., Maestas, E. G. M., & Darlington, J. D. (1999, June). Race, ethnicity and disasters in the United States: A review of the literature. *Disasters*, *25*(2), 156–173.

Godschalk, D. R., Beatley, T., Berke, P., Brower, D. J., Kaiser, E. J., Bohl, C. C., & Goebel, R. M. (Eds) (1999). *Natural hazard mitigation: Recasting disaster policy and planning*. Washington, DC: Island Press.

GovExec.com (2002). (on-line magazine), http://www.govexec.com/features/1102/1102s2.htm

Haas, C. N. (2002). The role of risk analysis in understanding bioterrorism. *Risk Analysis*, *22*(4), 671–677.

Haimes, Y. Y., & Longstaff, T. (2002). The role of risk analysis in the protection of critical infrastructures against terrorism. *Risk Analysis*, *22*(3), 439–444.

Harvard School of Public Health (2003). Promoting human security as a national security objective, http://www.hsph.harvard.edu/hpcr/

Heiken, G. F. et al. (2000). Modeling cities: The Los Alamos urban security initiative. *Public Works Management and Policy*, *4*(3), 198–212.

Hewitt, K. (1997). *Regions of risk*. Harlow: Longman.

Intergovernmental Panel on Climate Change (1997). The regional impacts of climate change: An assessment of vulnerability. In: R. T. Watson, M. C. Zinyowera & R. H. Moss (Eds), *A Special Report of IPCC Working Group II*. Cambridge and New York: Cambridge University Press.

Intergovernmental Panel on Climate Change (2001). *Climate change 2001: Impacts, adaptations, and vulnerability*. Contributions of Working Group II to the Third Assessment. Cambridge: Cambridge University Press.

International Institute for Sustainable Development (2003). *Environment and security*. Winnipeg. http://www.iisd.org/natres/security/

Kasperson, J. X., & Kasperson, R. E. (Eds) (2001a). *International workshop on vulnerability and global environmental change: A workshop summary*. Stockholm: Stockholm Environment Institute. http://www.sei.se/risk/workshop.html

Kasperson, R. E., & Kasperson, J. X. (Eds) (2001b). *Global environmental risk*. Tokyo, New York, Paris and London: United Nations University Press and Earthscan. pp. 10–11.

Kunreuther, H. (2003a). The role of insurance in managing extreme events: Implications for terrorism coverage. *Risk Analysis*, *22*(3), 427–437.

Kunreuther, H. (2003b). Risk analysis and risk management in an uncertain world. *Risk Analysis*, *22*(4), 655–664.

Lave, L. (2002). Viewpoint: Risk analysis and the terrorism problem in two parts. *Risk Analysis*, *22*(3), 403.

Mileti, D. S. (Ed.) (1999). *Disasters by design: A reassessment of natural hazards in the United States*. Washington, DC: Joseph Henry Press.

Mitchell, J. K. (2002). Building research capacity to address terrorism. *Countering terrorism: Lessons learned from natural and technological disasters*. National Research Council, Washington, DC, February 28–March 1. http://www7.nationalacademies.org/ndr/mitchell_abstract.html and http://www7.nationalacademies.org/ndr/Mitchell_Presentation.pdf

Mitchell, J. K. (2003a). Urban vulnerability to terrorism as hazard. In: S. L. Cutter, D. Richardson & T. J. Wilbanks (Eds), *The Geographical Dimensions of Terrorism* (pp. 17–25). New York: Routledge.

Mitchell, J. K. (2003b). *Urban disasters as indicators of global environmental change: Assessing functional varieties of urban vulnerability, Earth system science in the anthropocene*. Berlin: Springer-Verlag (forthcoming).

Moller, B. (1993). Security concepts: New challenges and risks, Working Paper 18. Copenhagen: Centre for Peace and Conflict Research.

National Research Council, Committee on Science and Technology for Countering Terrorism (2002). *Making the nation safer: The role of science and technology in countering terrorism*. Washington, DC.

National Research Council. Natural Disasters Roundtable (NDR) (2002). *Countering terrorism: Lessons learned from natural and technological disasters*, http://bob.nap.edu/books/NI000412/html/

Organization for Economic Cooperation and Development (2000). *State of the art review of environment, security and development cooperation*. Paris.

Pate-Cornell, E. (2002). Fusion of intelligence information: A Bayesian approach. *Risk Analysis*, *22*(3), 445–454.

Pew Research Center (2002). *What the world thinks in 2002: How global publics view – their lives, their countries, the world, America*. Washington, DC. Report 1652. pdf.

Platt, R. H. et al. (1999). *Disaster and democracy and disasters: The politics of extreme natural events*. Washington, DC: Island Press.

Popkin, R. (1990). The history and politics of disaster management in the United States. In: A. Kirby (Ed.), *Nothing to Fear: Risks and Hazards in American Society* (pp. 101–130). Tucson: University of Arizona Press.

Schwing, R. (2003). A mental model proposed to address sustainability and terrorism issues. *Risk Analysis*, *22*(3), 415–420.

Slovic, P. (2002). Terrorism as hazard: A new species of trouble. *Risk Analysis*, *22*(3), 425–426.

Wisner, B. (2002). Is there a geographical theory of terror? Panel Presentation at the Annual Meeting of the Association of American Geographers, Los Angeles, 19–24 March, 2002. http://www.anglia.ac.uk/geography/radix/resources/geographica-theory-of-terror-2002.doc

Witt, J. L., & Morgan, J. (2002). *Stronger in the broken places: Nine lessons for turning crisis into triumph*. New York: Times Books.

TERRORISM AS DISASTER: SELECTED COMMONALITIES AND LONG-TERM RECOVERY FOR 9/11 SURVIVORS

Brent K. Marshall, J. Steven Picou and Duane A. Gill

ABSTRACT

The purpose of this article is to apply what social scientists have learned from decades of research on natural and technological disasters to better understand the short-term and potential long-term human impacts of the 9/11 attacks. The short-term response to the 9/11 attacks was similar to how people and communities typically respond to natural disasters. One year after the attacks, news reports suggest that factors identified in technological disaster research as causing collective trauma, rather than recovery, are beginning to surface. We identify three patterns typically present in (but not restricted to) the aftermath of technological disasters that contribute to slow recovery and ongoing collective trauma and evaluate the likelihood that these factors will impact the recovery process for those impacted by the 9/11 attacks. We conclude that due to perceptions of governmental failure, the likelihood of protracted litigation, and uncertainty regarding the mental and physical health of victims, the social and psychological impacts of the 9/11 attacks will likely be severe and long-term. As such, the concluding section recommends the implementation of a long-term clinical intervention program for mitigating these potential chronic impacts and facilitating the timely recovery of survivors.

Terrorism and Disaster: New Threats, New Ideas
Research in Social Problems and Public Policy, Volume 11, 73–96

ISSN: 0196-1152/doi:10.1016/S0196-1152(03)11006-X

INTRODUCTION

It is increasingly apparent that large-scale disasters will be central features of the 21st century. These events pose significant problems for vulnerable populations and challenges for organizations responsible for disaster management. The 2002 *World Disasters Report* reveals that the impact of disasters has changed considerably over the last three decades, with the number of deaths decreasing and the number of those affected increasing (IFRC/RCS, 2002). For example, one million people died from disasters during the decade of 1982–1991, while there were 620,000 deaths from 1992 to 2001. This decline was largely due to an enormous drop in famine deaths in Africa. During the same two decades, the number of people affected by disasters increased from 1.7 to 2 billion. During 2001 alone, 170 million people were affected by disasters. People who are affected by disasters require "immediate assistance during a period of emergency, i.e. requiring basic survival needs such as food, water, shelter sanitation and immediate medical assistance" (IFRC/RCS, 2002, p. 181).

The United States has been substantially impacted by disasters over the last two decades. During the decades of 1982–1991 and 1992–2001, the total number of disaster fatalities increased slightly from 4,961 to 5,401, whereas the total number of people affected by disasters increased substantially from 229 thousand to 2.8 million. Compared to the 1970s, the direct economic loss due to all disasters increased dramatically to 629 billion U.S. dollars in the 1990s. Natural disasters such as the Northridge earthquake in California (1994) and Hurricane Andrew in South Florida and Louisiana (1992) were estimated to incur losses of $30 billion each. The U.S. has also experienced several large-scale technological disasters (e.g. Three Mile Island, *Exxon Valdez* Oil Spill), as well as chronically polluted areas (e.g. Love Canal (NY), Cancer Alley (LA)).

More recently, a third type of large-scale disaster has impacted the U.S. in ways that we as a nation and as individuals are still struggling to comprehend. Waugh, one of the first social scientists to address terrorism from an emergency management perspective, proclaimed seventeen years ago that "[n]o industrialized nation has to deal with actual acts of large-scale destruction outside of war. However, notwithstanding that good fortune, no one doubts that contemporary terrorists possess the technological capacities to commit such acts" (1986, p. 287). Unfortunately, Waugh's statement has proved prophetic. The bombing of the Alfred P. Murrah Federal Building in Oklahoma City in 1995 and the attacks on the Pentagon and World Trade Center (WTC) on September 11, 2001 have profoundly affected the American public. Given the gravity of these attacks and how little is known about the social impacts of terrorism, scholars and practitioners alike are challenged to better understand terrorism and its consequences.

The main purpose of this article is to apply what social scientists have learned from decades of research on natural and technological disasters to better understand the short-term and potential long-term human impacts of the 9/11 attacks. This article is divided into three sections. In the first section, we provide a brief historical sketch of disaster research in the United States. In the second section, we agree with existing research that the short-term response to the 9/11 attacks was similar to how people and communities typically respond to natural disasters. We then identify three patterns typically present in (but not restricted to) the aftermath of technological disasters that contribute to slow recovery and ongoing collective trauma and evaluate the likelihood that these factors will impact the recovery process for those impacted by the 9/11 attacks. Based on this evaluation, we conclude that due to perceptions of governmental failure, the likelihood of protracted litigation, and uncertainty regarding the mental and physical health of victims, the social and psychological impacts of the 9/11 attacks will likely be severe and long-term. As such, the concluding section presents a long-term clinical intervention program for mitigating these potential chronic impacts and facilitating the timely recovery of survivors.

DISASTER RESEARCH

The systematic study of disasters began in the late 1940s with researchers distinguishing between types of disasters mostly in physical terms (see, e.g. Lemons, 1957). Subsequent researchers defined an event as a disaster based on the physical consequences of the disaster, the way in which the impact of the disaster is evaluated, and the social effects of the disaster (Dynes, 1974). Reflecting this latter conception, Fritz (1961) defined a disaster as an event in which the "social structure is disrupted and the fulfillment of all or some of the essential functions of the society is prevented" (1961, p. 655). Much of the sociological research on disasters in the 1960s and 1970s studied the short-term, immediate social response to natural disasters (Fritz, 1961; Kreps, 1984; Quarantelli & Dynes, 1977).

These natural disaster researchers, based on a social constructionist perspective, argue that the consequences of any crisis situation can be viewed in terms of social structural vulnerabilities and the social definitions provided by people, organizations and communities (Kreps, 1989, 1998; Quarantelli, 1998). Thus, disaster characteristics are meaningless; an oil spill, tornado, earthquake and radiation release can all be appropriately labeled as a disaster, if the behavioral patterns of those impacted fit certain criteria defined by the research team. According to this perspective, the task of the researcher is to study victim response in the aftermath of each disaster *as if* the specific characteristics of the disaster does not affect the immediate impact and long-term response.

Early researchers noted the distinction between natural disasters as "acts of God" and technological disasters as "human-caused" (Barkun, 1974; Fritz, 1961). However, it was not until the emergence of large-scale, human-caused disasters in the late 1960s and 1970s that the social and psychological relevance of this distinction became a point of contention and the source of much debate among disaster researchers. Technological disaster researchers have found consistent empirical evidence that technological disasters create a far more exacting and enduring pattern of social, psychological, economic, and cultural impacts than do natural disasters (for a review, see Freudenburg, 1997). Natural disaster researchers have suggested that this distinction has been overemphasized and has little relevance to understanding disaster consequences (Alexander, 1993; Quarantelli, 1991, 1992, 1998).

We contend that research efforts to understand disaster impacts based solely on their physical properties (causes) or on the social responses (effects) underspecifies the critical relationship between specific causes and effects. People do not exist independently of their physical environment and disasters do not exist independently of subjective interpretation (Kroll-Smith & Couch, 1991). In other words, between the individual and the disaster is a "social process set in motion by the physical attributes of the agent itself and the symbolic capacities of humans" (Kroll-Smith & Couch, 1991, p. 361). The task of the researcher is to identify and distinguish between post-disaster factors that produce consensus-type crises and therapeutic processes and those factors that bring about conflict and corrosive processes. Only by more fully specifying these cause-effect relationships can we develop a long-term clinical intervention program that simultaneously sustains therapeutic processes while attenuating corrosive processes.

Although the natural and technological disaster research traditions are important and structure our efforts to assess the impacts of the 9/11 attacks, we make the broader claim that disaster researchers should shed the baggage of the natural-technological disaster debate and employ less encumbered approaches. The rigid conceptual and methodological distinction between natural and technological disasters, perhaps cogent when technological disasters first emerged, appears to be increasingly less meaningful. Although people may perceive some disasters as an act of God or nature, the severity and duration of disaster impacts may be ascribed to anthropogenic factors (see, e.g. Blocker & Sherkat, 1992; Erikson, 1976). Further, we assume there are post-disaster patterns unique to terrorist attacks that will have lasting social and psychological impacts that will be difficult to understand, let alone predict, from a natural or technological disaster perspective. For instance, terrorism is defined as a ". . . political, goal-directed action, involving the use or threat of extraordinary violence, performed for psychological rather than material effect, and the victims of which are symbolic rather than instrumental" (Waugh, 1986, p. 289). From this definition, it is clear terrorism is qualitatively different from natural and technological disasters. Another unique factor specific to the 9/11 attacks is the sheer number of fatalities, ranking it as one of the worst single-day death tolls in American history (WABF, 2002).

In short, we suggest that an *a priori* categorization of an event as either a natural or technological disaster is possible analytically, but increasingly difficult in the real world and is counterproductive. Therapeutic factors identified by natural disaster researchers and corrosive factors identified by technological disaster researchers may be present in the aftermath of any disaster situation. In the next section, we pose the following question: What post-disaster factors are present and emergent in the aftermath of the 9/11 attacks and will this mix of factors impede and/or enable long-term recovery?

THE 9/11 ATTACKS AS DISASTER

With the exception of fatalities, most of the damage stemming from a natural disaster is the destruction of built and modified environments, while technological disasters primarily damage the biophysical environment (e.g. the health of humans, animals, plants, and ecosystems) (Erikson, 1994). The impact of the 9/11 attacks on the built environment was devastating, both symbolically and materially. Rubble from the collapse of the WTC towers covered 16 acres, with an estimated weight of 1.2 million tons (WABF, 2002). Webb (2002) states that the "outpouring of support in response to large-scale [natural] disasters is often so great that the convergence of volunteers, supplies, and excessive donations on the scene of the disaster creates management problems" (p. 88). More broadly, Americans were "shocked into collective solidarity" due to televised images of "physical devastation, human misery, and heroic acts by civil servants and civilians" (Turkel, 2002, p. 73). Responses by New Yorkers, volunteers from within and outside the city, and the government seem to warrant the comparison to a natural disaster and it appears that something akin to a therapeutic community did emerge in lower Manhattan, and across America. However, this response may have been fleeting and the therapeutic community may be dissipating. One year after the attacks, news reports suggested that factors identified in technological disaster research as causing collective trauma, rather than recovery, were beginning to surface.

Research indicates that certain factors, typically present in the aftermath of technological disasters, have been found to produce a "corrosive community," one that impedes recovery (Freudenburg, 1993, 1997, 2000). The term "corrosive" implies that the negative impacts of some disasters damage a community and individuals over an extended period of time (Freudenburg, 1997). Victims do not follow the four-phase temporal trajectory of recovery identified by natural disaster researchers, but become arrested at an earlier corrosive stage (Couch, 1996). The question that should be addressed is what are the characteristics of technological disasters that preclude timely recovery and are such characteristics present in the post-9/11 attacks? For the sake of parsimony, we focus on three interrelated characteristics of the technological disasters: (1) the role of government and recreancy; (2) the consequences of litigation; and (3) mental and physical health problems. An overview of these factors will inform our discussion of the potential long-term impacts of the 9/11 attacks and highlight the necessity of developing a long-term clinical intervention program.

The Role of Government

In the aftermath of any disaster, the government's role and response are partially determined by disaster characteristics that influence the timeliness of recovery and whether or not collective trauma ensues. Governments in liberal democratic nations, such as the United States, "must fulfill two basic and often conflicting functions – accumulation and legitimization" (O'Connor, 1973, p. 6). Conditions conducive to economic growth are promoted via the accumulation function of the state because failure to do so undermines "the source of its own power, the taxes drawn from the economy's surplus production" (O'Connor, 1973, p. 6). Conversely, governmental actions (or inactions) that enable one class to accumulate wealth at the expense of other classes attenuate the legitimacy of the state, undermining the basis of its democratic support (O'Connor, 1973). Pressure on the state to fulfill its accumulation, not its legitimacy function, is pervasive, especially considering that with globalization, the power of governments has been decreasing relative to the increasing power of transnational corporations (Marshall, 1999, 2001; Robinson, 1996).

Given this power shift, governments have been pressured to adopt neoliberal policies, such as providing corporate subsidies and tax breaks and dismantling or not enforcing environmental regulations (Marshall, 1999, 2001). As certain segments of the population disproportionately face material deprivation and environmental degradation, and are increasingly vulnerable to disasters, the government's failure to fulfill its legitimacy function is revealed (Marshall, 2001). This revelation by an already apathetic citizenry further erodes confidence and trust in government. Citizens come to view particular governmental agencies as "recreant" – that is, they perceive institutional actors as failing to "carry out their responsibilities with the vigor necessary to merit the societal trust they enjoy" (Freudenburg, 2000, p. 108). Recreant behavior leads people "to believe that some portion of our increasingly complex and interdependent societal system can no longer be counted on" (Freudenburg, 1993, p. 927). Litigation further exposes experts, specialized organizations, and responsible parties for being irresponsible, incompetent and untrustworthy. Litigation exposes recreant behavior, thus contributing to the persistence of chronic disaster impacts through the loss of trust in traditional institutional support systems (Freudenburg, 1997).

In the aftermath of a disaster, victims will likely perceive government as recreant and untrustworthy if it fulfills its accumulation function, but not its legitimacy function. In turn, perceived recreancy and the loss of trust in government, often intensified by litigation, are likely to lead to collective trauma. If a disaster is viewed as an "act of God" or nature, then government's response is immediate and unequivocal. If the President declares an emergency, the Federal Emergency Management Agency (FEMA) quickly responds by providing temporary housing, emergency medical care, food supplies, clothes and other items essential for survival. While expanding its scope over years to include technological disasters and now terrorism, FEMA clearly has more experience with natural disasters. Government fulfills its legitimacy function by addressing the needs of the victims and, in some cases, fulfills its accumulation function by rebuilding an economically damaged area.

However, when disaster impacts are perceived as human-caused, the role of government becomes much more complicated and ambiguous. In these contested situations, the contradictory dual role of government, stimulating economic growth (corporate support) and maintaining legitimacy by meeting the needs of the people (victims), is magnified. The public may hold government responsible for ineffective disaster mitigation, an inadequate post-disaster response, and lackluster efforts to penalize the responsible party. Such public response is reflected in grassroots environmental movements (see, e.g. Cable & Cable, 1995) and the environmental justice movement (see, e.g. Bullard, 1990). Research on both movements provide evidence of government failure to fulfill its legitimacy function by not meeting the needs of victims of technological disasters, who are disproportionately the poor and people of color. Researchers have found that recreancy and mistrust of government are related to high levels of psychological distrust (Couch, 1996), perceived community damage (Picou et al., 2004), and heightened risk perceptions (Freudenburg, 1993; Marshall, 1995).

The role of government is even more complex when it comes to terrorism. Terrorism can be categorized as domestic or international depending on the origin of responsible parties. The 1995 bombing of the Murrah Federal Building in Oklahoma City was an act of domestic terrorism, but the attacks of 9/11 were international in origin. At least in one respect, the role of government is clear. The public expects the Federal government to protect citizens from external threats. Thus, unlike with technological disasters, government does not need to justify spending whatever resources are necessary to identify and capture the principle responsible party – the terrorists responsible for the disaster. As some proclaim, the 9/11 attacks were not just attacks on American soil, but more poignantly a symbolic attack on freedom writ large.

After the 9/11 attacks, many observers were puzzled that trust in government increased for the first time in three decades (Palmer & Samples, 2002). However, it appears that the National Election Studies scale may be too broad to capture how citizens view the complex role of government. In a recent ABC poll of the American public, it was found that "68% said they trust the government to do what is right when it comes to handling national security and the war on terrorism, compared to 38% who indicated the same level of trust when it comes to handling social issues . . ." (Palmer & Samples, 2002, p. 11). This finding is counterintuitive given that the public could have blamed the Federal government for not preventing the terrorist attack. Trust in government will likely decrease unless there is unequivocal proof that the terrorists, especially Osama bin Laden, have been brought to justice.

Clearly, government responded swiftly to fulfill its accumulation function. The WTC was the pinnacle of neoliberalism, symbolizing the virtues of private property, the accumulation of wealth, and the ascendancy of global capitalism (Turkel, 2002). The financial and political elite have not experienced such devastation and trauma since the sinking of the *Titanic* (Flynn, 2002). Flynn observed that "[w]hile the working class of America dug out bodies and mobilized for war, the wealthy class received what may become the largest financial bailout the world has ever seen" (2002, p. 105). Within two days of the 9/11 attacks, the Federal Reserve moved quickly to prevent a financial crisis, infusing an unprecedented $200 billion into the economy (Flynn, 2002). Furthermore, the Economic Security and Recovery Act of 2001 (HR 3090) was passed by the House of Representatives on October 24, 2001. While this bill only provided temporary tax relief to low-income households, massive tax cuts were given to the largest corporations, the minimum tax on corporations was eliminated, and income rates for high-income earners were permanently reduced (Turkel, 2002). The total financial aid package designed during the first week topped $300 billion, a public debt ultimately paid for by American taxpayers (Flynn, 2002).

Trust in government will likely decline as two potential sources of perceived recreant behavior by government are increasingly presented in the news. The first source concerns whether or not the intelligence community – the Central Intelligence Agency, the Federal Bureau of Investigation, and the National Security

Agency – is culpable for misreading warning signs and thus not preventing the 9/11 attacks. Further, the "appearance of guilt" may increase as the Bush administration stonewalls Congressional efforts to investigate governmental decision-making leading up to the 9/11 attacks. For example, the Administration prevented Secretary of State Colin Powell and Secretary of Defense Donald Rumsfield from testifying in public before the committee (Risen, 2002). Some lawmakers accused the Bush administration of "secretly trying to derail creation of an independent commission to investigate the Sept. 11 terrorist attacks while professing to support the idea" (Dewar, 2002, p. 1). This defiance underscores the strain between Congress and the President over the September 11 investigation, which may potentially embarrass the administration over its failure to prevent the attacks (Alden, 2002).

Another source of perceived recreancy is the role FEMA has played since the attacks. FEMA is a governmental agency which typically fills a positive rebuilding role after a natural disaster, but its role since 9/11 has been controversial. Due to numerous complaints, the agency revamped its programs in June 2002 by easing eligibility criteria for qualifying for the Mortgage and Rental Assistance Program, producing brochures in several languages, and speeding up the process of granting money to eligible applicants (Chen, 2002a). Many are complaining that FEMA is not distributing aid fairly. Recently, hundreds of people and community groups protested in Washington to demand that FEMA provide better treatment for low-income workers (Chen, 2002a). Since the attack, most of the federal aid has been used to cleanup and rebuild Ground Zero and assist major corporations displaced by the devastation (Gonzalez, 2002). A disproportionate amount of FEMA aid has gone to more affluent neighborhoods rather than poor neighborhoods, such as Chinatown and the Lower East Side (Chen, 2002c; Gonzalez, 2002). Carolyn B. Maloney, the Democratic Representative of Manhattan, summed up the feelings of many of her constituents stating that "[t]housands of people have lost trust in the agency because of the prior rejections and false promises the first time around" (as quoted in Chen, 2002a, p. 1). Compounding FEMA's ineffective response, the financial assistance program in New York State, known as the Individual and Family Grant Program, has also failed as a safety net (Chen, 2002b).

Emerging patterns suggest that the government may be viewed as not fulfilling its legitimacy function when it attempts to block lawsuits against numerous parties deemed indirectly responsible for the impacts of the 9/11 attacks. Further, if governmental toxicologists and epidemiologists continue to deny health claims made by lower Manhattan residents and volunteer workers, trust in government will likely decline. Although the government moved quickly to fulfill its accumulation function, literally within days of the attack, its efforts to fulfill its legitimacy function have taken more than a year and are perceived as inadequate by many victims.

Litigation

Litigation rarely ensues in the aftermath of natural disasters. However, as we have argued, if people come to perceive the severity and duration of disaster impacts as directly or indirectly rooted in anthropogenic forces, then litigation may become a more salient feature of the natural disaster recovery processes. To date, litigation typically has been a feature of technological disasters, the impacts of which have rarely been studied in the sociological literature. After a technological disaster occurs, victims of toxic contamination struggle to draw attention to their affliction, to attribute responsibility to an unresponsive government and/or evasive corporations, and ultimately to receive adequate redress in the legal system (Gill & Picou, 1991; Picou, 1996a, b; Picou & Rosebrook, 1993). Citizens today face a paradoxical situation:

> . . . at the very time when threats and hazards seem to become more dangerous and more obvious, they simultaneously slip through the net of proofs, attributions and compensation that the legal and political systems attempt to capture them with . . . [T]he responsibility to demonstrate proof, [Ulrich] Beck argues, currently lies overwhelmingly with the afflicted rather than potential polluters. The legacy of industrial society's faith in progress is that the legal system assumes that industrial production will be benign unless demonstrated otherwise (Goldblatt, 1996, pp. 166–167).

Once litigation begins, the principal responsible party has an inherent advantage because the legal system "demands proofs of post hoc toxicity rather than pre hoc non-toxicity or safety" and victims who must prove toxicity are inevitably less endowed with the resources necessary to make a convincing case (Goldblatt, 1996, p. 167). The public generally does not wield the appropriate scientific discourse to establish adequate proof in demonstrating the cause-effect linkage necessary for legal redress. The linkage between a single substance and a particular set of pathological effects may be difficult to establish given the likelihood that symptoms of the afflicted are a function of the synergistic affect of exposure to multiple chemicals. Further, the public typically does not have the resources to employ a cadre of legal experts that lobby for corporate liability in the legal system. Information overload, although cooperative in a legal sense, seeks to obfuscate and bury officials with information that is irrelevant as evidence of code violations. Corporate claims that victims lack scientific knowledge undercut the legitimacy of the victims' criticism, thus maintaining technical hegemony and institutional legitimacy.

Furthermore, given the private nature of the discovery phase of litigation, as well as sealed legal settlements, systematic data on the consequences of litigation have rarely been collected and reported. Research has found that individuals and communities chronically impacted by technological disasters experience additional significant impacts from the litigation process. Impacts include conflict over equitable damage payments, stress from protracted legal procedures and the overall uncertainty of litigation outcomes. In short, litigation contributes to the inability of individuals and communities impacted by technological disasters to embark on a timely recovery process, resulting in long-term collective trauma (Picou et al., 2004).

Although litigation primarily has been a feature of technological disasters in the United States, we suggest that a protracted litigation process will have similar corrosive effects for post-disaster recovery in any disaster situation. Although it is too early to tell if litigation will emerge as a long-term outcome of the 9/11 attacks, news reports indicate that trial lawyers are preparing for lawsuits. It should be noted that, unlike the situation in technological disasters, it would be rare if the principal party directly responsible for the disaster (the terrorists) will be sued. However, those perceived as indirectly responsible for the impacts of the 9/11 attacks will likely bear the burden of litigation. For instance, the air transportation industry, particularly airport security, received most of the blame for the terrorist attacks. National polls conducted days after the attacks found that 78% of Americans blamed airport security for the terrorists attacks, two-thirds believed that the attacks could have been prevented with tighter airport security (CBS News/*New York Times*, 2001), and 82% thought that airports in the United States were doing too little to secure the safety of passengers (*LA Times*, 2001).

The government also received blame by Americans just days after September 11, especially the U.S. intelligence community for allowing the attacks to occur. For instance, 56% of Americans felt that U.S. intelligence should have known about the attacks in advance (CBS News/*New York Times*, 2001), 51% agreed that government intelligence agencies should have done more to prevent the attacks, and 58% directly linked the attacks with U.S. policies in the Middle East (*LA Times*, 2001). Specifically, the CNN/USA Today/Gallup (2001) poll conducted September 14–15 found that 59% blamed the CIA and 52% blamed the FBI for not preventing the attacks. Incongruous with the statistics above, the Ipsos-Reid (2001) poll reported that 88% had confidence in U.S. military defense and 62% expressed confidence in the FBI. The September 11th CBS News/*New York Times* (2001) poll found that two-thirds expressed confidence in the ability of the U.S. government to prevent further terrorist attacks against America and more than 90% expressed confidence in the ability of the U.S. government to find and punish the people responsible for the attacks.

How do we explain the above findings indicating that the American public partially blames the U.S. intelligence agencies for not preventing the terrorist attacks, but conversely expressed confidence in the U.S. government to find and punish terrorists and prevent future attacks? One explanation, unique to issues of national security, is the "rally effect." Although blame was placed on U.S. intelligence agencies, confidence in government improved as citizens rallied behind political leaders in this time of crisis. President Bush's approval rating soared after the attacks. Three polls indicated that a large majority (64–78%) of Americans expressed confidence in and approval of the president's ability to handle the crisis (CBS News/*New York Times*, 2001; CNN/USA Today/Gallup, 2001; Ipsos-Reid, 2001). Another poll found that the president's approval rating jumped 35 percentage points from 51 to 86% (highest ever recorded by Gallup). Furthermore, 91% approved of the way the president was handling the events surrounding the terrorist attacks and the approval of Congress increased from 42% prior to the attacks to 75% (CNN/USA Today/Gallup, 2001). This is the highest approval rating ever reported by a national poll for Congress. The rally effect and time needed to grieve may have delayed widespread litigation.

Although litigation has not been a central feature in the aftermath of the 9/11 attacks, lawsuits are currently being filed and there is evidence that more lawsuits will follow. This delay is due, in part, to active steps taken by Congress and the Bush administration to avoid a "class-action free-for-all" (Carney & Walczak, 2001, p. 114). By capping liability at $6 billion (the amount the airlines were insured for), the Administration was able to protect United Airlines and American Airlines from immediate bankruptcy. The White House has introduced legislation to protect other companies, such as WTC property manager Larry Silverstein and the Boeing Company (Carney & Walczak, 2001).

Congress also acted by creating the government-supported Victims Compensation Fund (VCF) to prevent litigation and provide reliable and expeditious compensation to victims' families in order to spare them the uncertainties of drawn-out litigation ("Warily Circling," 2002). Kenneth Feinberg (the director of the VCF appointed by the Bush administration) has announced criteria for estimating each compensation package. The family of "each survivor will receive $250,000 for pain and suffering, plus $50,000 per child, plus an amount based on the victim's age and income" (Weinbaum & Page, 2002, p. 1). Low-income families, high-income families, uniformed rescue workers, and other groups contend that, for different reasons, they will not receive the compensation package they deserve (Paulson, 2002). Even though Feinberg estimates the average payout will be $1.5 million, only 10 families out of 3,200 claimants have forfeited the right to sue by filing completed applications ("Warily Circling," 2002). At the same time, trial lawyers and clients are organizing, preparing, and filing lawsuits.

Nearly 3,000 relatives of 9/11 victims have filed lawsuits against three Saudi princes, seven international banks (mainly in the Middle East), eight Islamic foundations and charities, and the government of Sudan for allegedly funding Osama bin Laden and Al Qaeda ("Sept. 11 Survivors," 2002). In response to reports that the Bush administration may block or delay their lawsuit, many relatives held a protest rally at the U.S. Capitol (Mintz, 2002). In addition, more than 1,000 families have filed legal papers preserving their right to sue the Port Authority of New York and New Jersey for lapses in safety and poor evacuation strategies at the WTC (Weiser, 2002). Approximately 1,050 firefighters, afraid they won't be eligible for the VCF, have filed a notice of intention to sue New York City, most likely for not providing the proper equipment to protect against toxic substances at the disaster site (Purnick, 2001). These firefighters and other rescue workers claim that breathing toxic substances thrown in the air by the collapse and cleanup of the WTC have made them ill (Van Voris, 2002). Lawyers predict that lawsuits filed thus far are the tip of the iceberg.

The critical question is what happens after the "rally effect" subsides and it is no longer unpatriotic for people to question the government's culpability in not preventing the disaster, or to sue those indirectly responsible for loss of life and emotional damages. It appears that widespread litigation is on the horizon. As such, it is likely that those directly affected by the 9/11 attacks will be facing an adversarial, protracted, litigation process which leads to chronic psychosocial stress rather than recovery.

Mental and Physical Health Issues

Disaster survivors are potentially faced with two types of health issues – mental and physical. Evidence of Posttraumatic Stress Disorder (PTSD), although not officially recognized as a psychiatric disorder until 1980 (APA, 1980), was found in studies of Vietnam war veterans. PTSD is an anxiety disorder that can develop after exposure to a traumatic event or ordeal – such as, violent personal assaults, natural or human-caused disasters, accidents, or military combat – in which grave physical harm occurred or was threatened (NIMH, 2002). Other psychological problems that often accompany PTSD include emotional numbness, sleep disturbances, depression, anxiety, irritability, outbursts of anger, and feelings of intense guilt. Natural disasters, technological disasters, and terrorism can cause PTSD and other psychological problems. Nonetheless, PTSD symptoms have been highly characteristic of victims of technological disasters (Arata et al., 2000; Baum & Fleming, 1993; Green, 1996; Picou & Gill, 1996, 2000).

In a comprehensive review of disaster studies, Norris et al. (2001) evaluated 177 articles which analyzed the psychological effects of natural and human-caused disasters on 130 distinct samples of victims. Of the disasters studied, 62% were natural disasters, 29% technological disasters, and 9% mass violence. Specific psychological problems and PTSD were found in 74% and 65% of the 130 distinct samples of victims, respectively. These researchers found that, in the U.S., technological disasters were more psychologically stressful than natural disasters. However, disasters of mass violence were the most stressful of all three types. Despite the apparent relationship between disaster type and psychological stress, evidence indicates that extremely damaging natural disasters (e.g. Hurricane Andrew in 1992), technological disasters (e.g. the 1972 dam collapse in Buffalo Creek, WV and the 1989 *Exxon Valdez* oil spill in Alaska), and disasters of mass violence (e.g. the 1995 bombing in Oklahoma City and, undoubtedly, the 9/11 attacks) can cause severe to very severe psychological impairment (Norris et al., 2001). The authors conclude that psychological impacts were severe, lasting and pervasive when at least two of the following factors were present: (1) extreme and widespread damage to property; (2) serious and ongoing financial problems for the community; (3) human carelessness or, especially, human intent; and (4) high prevalence of trauma in the form of injuries, threats to life, and loss of life. Given that all four of the factors are present in the aftermath of the 9/11 attacks, a long-term clinical intervention program is needed to minimize the severity and duration of the psychological impacts on victims of 9/11.

As noted, disasters of mass violence, such as terrorism, cause more severe psychological impairment than other types of disasters (Norris et al., 2001). For instance, six months after the Oklahoma City bombing, 34% of the survivors were suffering from PTSD (North et al., 1999). Two years later, 16% of children and youth living up to 100 miles from Oklahoma City exhibited symptoms of PTSD in relation to the bombing (Pfefferbaum et al., 1999). Initial studies on the 9/11 attacks seem to support this pattern. In a survey of adults living south of 110th Street in Manhattan five to eight weeks after September 11, researchers found that 7.5% and 9.7% of the respondents reported symptoms consistent with the diagnosis of current PTSD and depression, respectively (Galea et al., 2002). Based on comparisons to national studies, Galea et al. (2002) suggest that the prevalence of PTSD and depression in lower Manhattan were twice as high as expected. The researchers found significant relationships between low levels of social support and symptoms of PTSD and depression (Galea et al., 2002). This pattern is consistent with long-term studies of victims of the Three Mile Island radiation release (Baum & Fleming, 1993) and the *Exxon Valdez* oil spill (Arata et al., 2000).

What is unique about the 9/11 attacks, and perhaps terrorism more generally, is the degree of trauma experienced nationally by those not directly impacted by the disaster. For example, Schuster et al. (2001) conducted a national telephone survey of 560 adults three to five days after the 9/11 attacks. They found that 44% reported one or more symptoms of stress, including PTSD, and 90% had one or more symptoms to at least some degree. Compared to others parts of the U.S., the incidence of stress was greater in the Northeast region (55% reported substantial stress) and within a 100 mile radius of the WTC (61% reported substantial stress) (Schuster et al., 2001). In a national telephone survey of adults conducted three months after the 9/11 attacks, Gill et al. (2002) found that 30% reported one or more substantial symptoms of stress, including PTSD. Further, almost 40% in the Northeast region of the U.S. reported substantial symptoms of stress (Gill et al., 2002). In a 3-panel Web-based design of the U.S. population outside of New York City, Silver et al. (2002) found that 17% of the population reported PTSD two months after the attacks. Although stress levels had declined six months later, 6% were still elevated and higher than normal. This study belies the unwarranted assumption "that subjective responses to trauma are proportional to the degree of objective loss experienced" (Silver et al., 2002, p. 1235).

Assessing the long-term effects of the 9/11 attacks is premature, especially considering that many victims may exhibit "serious symptoms of distress without qualifying for a clinical diagnosis of PTSD but later move into this diagnostic category, sometimes with a delayed onset of nearly 10 years" (Kaspar, 2002, p. 100). Another reason why it is premature is that secondary stress producing factors, such as recreancy, litigation, and physical health problems may prolong and exacerbate existing levels of psychological stress.

The physical health problems stemming from hurricanes and earthquakes typically result in fatalities and injuries to many victims in high impact areas. However, when people and communities are exposed to toxic chemicals in the aftermath of a disaster, the most troubling outcomes has to do not with direct loss of life, but the potential for long-term health problems and damage to the community. More specifically, toxic chemicals are often imperceptible to the senses and victims are uncertain about the nature and extent of their exposure to toxins (Erikson, 1994). The effects of toxic pollution are usually chronic and accumulative, not acute. It is very difficult to identify the point(s) of contamination and impossible to predict whether the effects will abate. The physical symptoms of toxic contamination often mimic those of other ailments, which result in delayed diagnosis (Erikson, 1994). Further, the medical community is unable to estimate the immediate and future damage caused by exposure to toxins, obscuring the identification of victims from non-victims and producing a "contested" discourse concerning health impacts (Brown, 1997; Couch & Kroll-Smith, 1985; Kroll-Smith et al., 2002).

The Environmental Protection Agency's Christine Todd Whitman, seven days after the 9/11 attacks, stated, "[T]he public in these areas is not being exposed to excessive levels of asbestos or other harmful substances," and that "New Yorkers need not be concerned about environmental issues as they return to their homes and workplaces" (as quoted in Scelfo & Smalley, 2002, p. 4). Such assurances were based on an air-monitoring system, set up after the attack, designed to measure long-term exposures associated with normal air pollution or workplace air quality (Johnson, 2002a). However, no existing monitoring system could have accurately measured the impacts after the attacks, given that "[t]he blast of dust and smoke – and the toxic substances, fibers and ash that blew through New York in the days afterward – is without precedent in medical literature..." (Johnson, 2002a, p. 1). The potentially harmful substances in the air and dust in the neighborhood of the WTC include asbestos, lead, fiberglass, PCBs, mercury and other potentially harmful particulates (Van Voris, 2002). Six months after the 9/11 attacks, approximately 12% of the firefighters at ground zero developed the "World Trade Center cough," a severe respiratory problem requiring medical leave for at least four weeks (Prezant et al., 2002).

Uncertainties regarding health abound. For instance "some compounds that were created in the fires of the trade center... have never been seen in nature or in the laboratory" (Johnson, 2002b, p. 2). Even with reliable tests, health impacts are uncertain "because the government has not established standards for the amount of asbestos allowable in residential air" (Johnson, 2002b, p. 3). Much of the problem of assessing the impacts of the WTC collapse on air quality and respiratory illnesses is that baseline data are practically non-existent, especially in particular neighborhoods like Chinatown (Johnson, 2002b). Nearly 30 times the federally acceptable level of lead has been found at a downtown Manhattan high school, presumably contaminated by debris from the ground zero cleanup being transported past the school (Hortocollis, 2002).

Again, it is too early to tell if and to what degree the residents of lower Manhattan and volunteer workers will suffer long-term health problems due to exposure to airborne toxic substances. As indicated, journalists and residents are deeply concerned and uncertainty exacerbates the situation. Evidence from the technological disaster literature suggests that if people in lower Manhattan are sick (real or perceived) due to exposure to toxic chemicals, and scientists continue to deny the legitimacy of these claims, perceptions of recreancy are likely to increase, toxic lawsuits will ensue, and psychosocial stress levels will increase and be prolonged.

CONCLUSIONS AND IMPLICATIONS

In this article we have identified characteristics of catastrophic natural and technological disasters that are relevant to understanding the potential long-term social and psychological consequences of the 9/11 attacks. Initial responses to the 9/11 attacks have followed the traditional natural disaster model. That is, volunteers from within and outside the impact areas rushed to ground zero to lend a helping hand. Many firefighters and police officers went above and beyond the call of duty and countless heroic acts by ordinary people were highlighted on the nightly news. The survivors and the American public provided high-levels of support for government and expressed trust regarding the ability of the government to catch and punish the terrorists. The approval ratings of both Congress and the President jumped to record levels. However, this seemingly therapeutic response may be less reflective of the initial similarities between the 9/11 attacks and natural disasters per se, but more reflective of the rally effect that tends to accompany any threat to national security. Nearly 14 months later, the rally effect for victims of the 9/11 attacks seems to be dissipating.

Serious contested issues are being raised about: (1) the responsibility of the intelligence community for the 9/11 attacks; (2) the Bush administration's reluctance to cooperate with the Congressional 9/11 investigation; (3) governmental efforts to forestall the filing of class-action lawsuits against those deemed indirectly responsible for the attacks; (4) complaints regarding the perceived class-biased distribution of aid by FEMA; (5) the purpose and fairness of the proposed compensation package for survivors offered by Congress; (6) the potentially chronic respiratory illnesses experienced by rescue workers and residents of lower Manhattan; and (7) potential mental health problems. Distilled from this contested discourse, we have identified the issues of recreancy, litigation, and mental and physical health as having potentially long-term negative psychosocial impacts for survivors. As such, we argue for an expanded disaster framework that will allow an understanding of the future consequences of the 9/11 attacks. Over time, continuing social structural impacts may result in ineffective coping responses by survivors, which include, but are not limited to, loss of trust in government, litigation and the emergence of an ongoing, adversarial corrosive community (Freudenburg, 1997; Picou, 1996b; Picou & Gill, 2000).

Given that litigation and corrosive communities are most often associated with the chronic impacts of technological disasters, we briefly reviewed trends in the technological disaster literature regarding long-term social and psychological impacts. This literature clearly indicates that survivors experience anger, fear, uncertainty and psychological stress from exposure to toxic chemicals. Often such claims are denied legitimacy to survivors, further exacerbating their concerns and fears. Over time, chronic psychosocial problems emerge, including depression, anger, PTSD and anxiety (Arata et al., 2000; Baum & Fleming, 1993; Couch & Kroll-Smith, 1985; Green, 1996), placing survivors in a continuous "warning/threat/impact" cycle which is characteristic of technological disasters (Couch, 1996).

The relevance of the technological disaster literature for understanding the impacts on survivors of the 9/11 attacks relates specifically to the time dimension. Over time, victims of technological disasters often form groups based on common interests. Each group may define the situation differently, producing the seeds of intense and corrosive intracommunity conflict and making long-term recovery difficult (Couch, 1996). Indeed, "recovery and emergency response must take place simultaneously," making any clinical intervention strategy problematic (Couch, 1996, p. 74). Intracommunity conflict tends to be more socially and psychologically damaging when group definitional differences are vast and if outside interest groups get involved (Couch, 1996). We have presented preliminary evidence that such a conflict-based legal scenario may yet result from the 9/11 attacks. The "contested nature" of damage claims of survivors and the reaction of defendants may result in a legal disclosure that will be a daily reminder of the trauma experienced from the 9/11 attacks. This outcome will only add to the long-term negative impacts of survivors, reinforcing chronic psychological impacts, relational disorders, and ineffective coping strategies among victims (Arata et al., 2000; Picou et al., 2004).

Given the similarities between technological disasters and the 9/11 attacks, it is apparent that a long-term clinical intervention program should be developed and provided to survivors of the 9/11 attacks. There have been limited efforts to develop such programs, and several researchers have recently proposed concepts and methods for implementing mitigation strategies for communities and individuals chronically impacted by technological disasters (Couch, 1996, 1999; Picou, 2000; Picou et al., 2003; PWSRCAC, 1999). These efforts, when applied to the WTC attacks, provide one basis to establish a long-term clinical infrastructure to respond to the most prominent goal of international terrorism, i.e. the destruction of social support and the undermining of existing government legitimacy.

A Long-Term Clinical Program

At the most general level, a federally-funded task force should monitor the recovery of victims and provide ongoing recommendations for program characteristics and treatment options. Such a program could be developed under the guidance of the National Academy of Sciences, or a similar agency, and serve as an objective, independent advisor for the treatment of survivors (Couch, 1996). The management of survivor recovery should be centralized, with one agency (such as the National Institute of Mental Health) being responsible for programmatic intervention. Clear lines of authority should be established and the designated agency should be responsive to the independent task force identified above. The long-term clinical intervention program should, first and foremost, utilize a participatory methodology to empower survivors to transform themselves.

Participatory empowerment is a recognized strategy in the treatment of PTSD and has been successfully employed in a clinical intervention for mitigating the chronic (six years) social impact of a technological disaster, the *Exxon Valdez* Oil Spill (Picou, 2000; Picou et al., 2003; PWSRCAC, 1999). This program developed culturally sensitive treatment programs (Picou, 2000), utilized a participatory methodology, employed outreach strategies, focused on empirically diagnosed psychological symptoms and relational disorders (Arata et al., 2000), included educational training for lay peer-listeners, and provided extensive programmatic educational materials on coping strategies to victims (Picou et al., 2003; PWSRCAC, 1999). These programmatic efforts provide the foundation for the development of an effective counter-terrorism intervention model for survivors of the WTC attacks. More importantly, establishment of effective victim intervention programs provide a measure of counter-terrorism. Intervention programs can help reassure potential victims that they will be cared for, thus reasserting a level of security. Given the predicted increase in the number of terrorist acts in the 21st century, it is important that our society take a proactive stance at this level.

In conclusion, we are aware of the complexity and detail involved in developing a long-term clinical intervention program for mitigating the potential mental health consequences of the WTC attacks. Nonetheless, the literature on other human-caused disasters provides strong empirical evidence that long-term psychological and relational disorders may characterize the survivors of 9/11. The knowledge and understanding of such chronic disaster impacts are well-understood by clinical and applied psychologists, sociologists and anthropologists who have studied a wide-variety of disasters over the last forty years (Norris et al., 2001). We also suggest that rather than waiting for chronic symptomologies to emerge among 9/11 survivors, serious consideration be given by Federal authorities to initiate the discussion, development and implementation of a comprehensive long-term clinical intervention program based on the basic and applied social science literature on technological disaster impacts. Since the 9/11 attacks, Americans have been repeatedly reminded that the war on terrorism will be a long, drawn-out battle for the United States. In this article, we have attempted to point out that evidence from the social science literature on the chronic social impacts of technological disasters portends a similar long, drawn-out battle for social and psychological recovery for the survivors of this catastrophic act of terrorism.

NOTES

1. The technological disaster category includes transport accidents, miscellaneous accidents, and industrial accidents (IFRC/RCS, 2002). Environmental sociologists studying

the impact of technological disasters typically focus on industrial accidents. The impact of transport accidents is an understudied area in the social sciences.

2. Survivors will be used in this article in a broad manner to refer to individuals injured, families and friends of fatalities, rescue workers, and others who may have resided in the impact area of the WTC attacks.

ACKNOWLEDGMENTS

The comments provided by Bob Gossweiler, David Gartman, Nicole Flynn, and Christine Bevc on an earlier draft are gratefully acknowledged. Support for this work has been provided by the Department of Sociology and Anthropology at the University of Central Florida, National Science Foundation (OPP-0002572), and the Mississippi Agricultural and Forestry Experiment Station (MIS-605080). The authors are solely responsible for the contents. Direct correspondence to Dr. Brent K. Marshall, Department of Sociology and Anthropology, University of Central Florida, Orlando, FL 32816-1360. E-mail: bmarshal@pegasus.cc.ucf.edu

REFERENCES

Alden, E. (2002). White House defies inquiry over 9/11 intelligence. *Financial Times* (London), September 19, p. 8. Retrieved October 8, 2002, from LexisNexis™ Academic.

Alexander, D. (1993). *Natural disasters*. New York: Chapman and Hall.

American Psychiatric Association (1980). *Diagnostic and statistical manual of mental disorders* (3rd ed.). Washington, DC: American Psychiatric Association.

Arata, C. M., Picou, J. S., Johnson, G. D., & McNally, T. S. (2000). Coping with technological disaster: An application of the conservation of resources model to the *Exxon Valdez* oil spill. *Journal of Traumatic Stress*, *13*, 23–39.

Barkun, M. (1974). *Disaster and the millennium*. New Haven, CT: Yale University Press.

Baum, A., & Fleming, I. (1993). Implications of psychological research on stress and technological accidents. *American Psychologist*, *48*, 665–672.

Blocker, J., & Sherkat, D. (1992). In the eyes of the beholder: Technological and naturalistic interpretations of a disaster. *Industrial Crisis Quarterly*, *6*, 153–166.

Brown, P. (1997). Popular epidemiology revisited. *Current Sociology*, *45*, 137–156.

Bullard, R. D. (1990). *Dumping in Dixie: Race, class, and environmental quality*. Boulder, CO: Westview Press.

Cable, S., & Cable, C. (1995). *Environmental problems, grassroots solutions: The politics of grassroots environmental conflict*. New York, NY: St. Martins Press.

Carney, D., & Walczak, L. (2001). September 11: Here come the trial lawyers. *Business Week*, November 12, 3757, p. 114. Retrieved October 8, 2002, from LexisNexis™ Academic.

CBS News (2001). Poll data retrieved October 3, 2001, from http://www.cbsnews.com/new/section/0.1636.215-412.00.shtml

Chen, D. W. (2002a). More get 9/11 aid, but distrust of U.S. efforts lingers. *New York Times*, August 27, p. 1. Retrieved October 8, 2002, from LexisNexis™ Academic.
Chen, D. W. (2002b). Seen as safety net, 9/11 program is anything but. *New York Times*, September 28, p. 1. Retrieved October 8, 2002, from LexisNexis™ Academic.
Chen, D. W. (2002c). Public frustration persists over 9/11 relief program. *New York Times*, October 12, p. 3. Retrieved October 8, 2002, from LexisNexis™ Academic.
Couch, S. R. (1996). Environmental contamination, community transformation and the Centralia Mine fire. In: J. K. Mitchell (Ed.), *The Long Road to Recovery: Community Responses to Industrial Disaster* (pp. 60–84). New York: United Nations University Press.
Couch, S. R. (1999). Recovery from chronic technological disasters: Is it possible? Paper presented at the meeting of the Eastern Sociological Society, March, Boston, MA.
Couch, S. R., & Kroll-Smith, J. S. (1985). The chronic technical disaster: Toward a social scientific perspective. *Social Science Quarterly*, *66*, 564–575.
Dewar, H. (2002). White House accused of thwarting 9-11 panel. *The Washington Post*, October 12, p. 1. Retrieved November 1, 2002, from LexisNexis™ Academic.
Dynes, R. R. (1974). *Organized behavior in disaster*. Lexington, KY: Heath.
Erikson, K. (1976). *Everything in its path: Destruction of community in the Buffalo Creek flood*. New York: Simon and Schuster.
Erikson, K. (1994). *A new species of trouble*. New York: Norton.
Federal Emergency Management Agency (2002). Retrieved October 2, 2002, from http://www.fema.gov/
Flynn, P. (2002). Financial bailout of September 11: Rapid response. *Challenge*, January/February, pp. 104–116. Retrieved October 8, 2002, from LexisNexis™ Academic.
Freudenburg, W. R. (1993). Risk and recreancy: Weber, the division of labor, and the rationality of risk perceptions. *Social Forces*, *71*, 909–932.
Freudenburg, W. R. (1997). Contamination, corrision, and the social order: An overview. *Current Sociology*, *45*, 19–40.
Freudenburg, W. R. (2000). The 'Risk Society' reconsidered: Recreancy, the division of labor, and risk to the social fabric. In: M. J. Cohen (Ed.), *Risk in the Modern Age: Social Theory, Science and Environmental Decision Making* (pp. 107–120). New York: St. Martin's Press.
Fritz, C. E. (1961). Disaster. In: R. K. Merton & R. A. Nisbet (Eds), *Contemporary Social Problems* (pp. 651–694). New York: Harcourt.
Galea, S., Ahern, J., Resnick, H., Kilpatrick, D., Bucuvalas, M., Gold, J., & Vlahov, D. (2002). Psychological sequelae of the September 11 terrorist attacks in New York City. *The New England Journal of Medicine*, *346*, 982–987.
Gallup Organization (2001). Poll data retrieved September 18, 2001, from http://www.gallup.com
Gill, D. A., Parisi, A., Cosby, G., & Frese, W. (2002). Societal impacts of the September 11th terrorist attacks: general findings from a national survey. Paper presented at the annual meeting of the Alabama-Mississippi Sociological Association, Cleveland, MS.
Gill, D. A., & Picou, J. S. (1991). The social psychological impacts of a technological accident: Collective stress and perceived health risks. *Journal of Hazardous Materials*, *27*, 77–89.
Goldblatt, D. (1996). *Social theory and the environment*. Boulder, CO: Westview Press.
Gonzalez, J. (2002). Sept. 11 aid seems to elude the poor. *Daily News* (New York), June 6, p. 40. Retrieved October 8, 2002, from LexisNexis™ Academic.
Green, B. L. (1996). Traumatic stress and disaster: Mental health effects and factors influencing adaptation. In: F. L. Mak & C. C. Nadelson (Eds), *International Review Psychiatry* (Vol. 2). Washington, DC: American Psychiatric Press.

Hortocollis, A. (2002). Lead levels at school are linked to Sept. 11. *New York Times*, June 15, p. 3. Retrieved October 8, 2002, from LexisNexis™ Academic.

International Federation of Red Cross/Red Crescent Societies (2002). *World disasters report*. Bloomfield, CT: Kumarian Press.

Ipsos-Reid (2001). Poll data retrieved September 18, 2001, from http://www.ipsos-reid.com

Johnson, K. (2002a). Studies will take Sept. 11's measure in health effects. *New York Times*, January 11, p. 1. Retrieved October 8, 2002, from LexisNexis™ Academic.

Johnson, K. (2002b). Air testing after Sept. 11 attack is both perplexing and reassuring. *New York Times*, May 15, p. 1. Retrieved October 8, 2002, from LexisNexis™ Academic.

Kaspar, V. (2002). Posttraumatic stress disorder: Diagnosis, prevalence, and research advances. *Sociological Focus*, *35*, 97–108.

Kreps, G. (1984). Disaster and the social order. *Sociological Theory*, *3*, 49–65.

Kreps, G. (1989). *Social structure and disaster*. Newark, DE: University of Delaware Press.

Kreps, G. (1998). Disaster as systemic event and social catalyst. In: E. L. Quarantelli (Ed.), *What is Disaster*. New York: Routledge.

Kroll-Smith, J. S., & Couch, S. R. (1991). What is a disaster? An ecological-symbolic approach to resolving the definitional debate. *International Journal of Mass Emergencies and Disaster*, *9*, 355–366.

Kroll-Smith, J. S., Couch, S. R., & Levine, A. G. (2002). Technological hazards and disasters. In: R. E. Dunlap & W. Michelson (Eds), *Handbook of Environmental Sociology* (pp. 295–328). Westport, CT: Greenwood Press.

Lemons, H. (1957). Physical characteristics of disaster: Historical and statistical review. *Annals of the American Academy of Political and Social Science*, *309*, 1–14.

Los Angeles Times (2001). Poll data retrieved September 27, 2001, from http://www.latimes.com/news/custom/timespoll

Marshall, B. K. (1995). Vulnerability, environmental degradation and confidence in local government. Unpublished Masters Thesis, University of New Orleans, New Orleans.

Marshall, B. K. (1999). Globalization, environmental degradation, and Ulrich Beck's risk society. *Environmental Values*, *8*, 253–276.

Marshall, B. K. (2001). Bridges and barriers to ecosystem-based approaches: The case of the Tennessee Valley Authority's adoption of the watershed approach. Unpublished doctoral dissertation, University of Tennessee, Knoxville.

Mintz, J. (2002). 9/11 Relatives, Suing Saudis, to Protest Today. *The Washington Post*, November 1, p. 1. Retrieved November 2, 2002, from LexisNexis™ Academic.

National Institute of Mental Health (2002). Facts about post-traumatic stress disorder. Retrieved October 12, 2002, from http://www.nimh.nih.gov/anxiety/ptsdfacts.cfm

Norris, F., Byrne, C. M., Diaz, E., & Kaniasty, K. (2001). The range, magnitude, and duration of effects of natural and human-caused disasters: A review of the empirical literature. National Center for PTSD. Retrieved October 13, 2002, from http://www.ncptsd.org/facts/disasters/fs_range.html

North, C. S., Nixon, J. S., Shariat, S., Mallonee, S., McMillen, E., Spitzbagel, E. L., & Smith, E. M. (1999). Psychiatric disorders among survivors of Oklahoma city bombing. *JAMA*, *282*, 755–762.

O'Connor, J. (1973). *The fiscal crisis of the state*. New York: St. Martin's Press.

Palmer, T. G., & Samples, J. (2002). Limiting government after September 11. *USA Today*, July, pp. 10–12. Retrieved October 8, 2002, from LexisNexis™ Academic.

Paulson, A. (2002). Divvying up of 9/11 aid leaves few happy [electronic version]. *Christian Science Monitor*, January 14. Retrieved October 10, 2002, from http://www.csmonitor.com/2002/0114/p3s1-usgn.html

Pfefferbaum, B., Nixon, S. J., Krug, R. S., Tivis, R. D., Moore, V. L., Brown, J. M., Pynoos, R. S., Foy, D., & Gurwitch, R. H. (1999). Clinical needs assessment of middle and high school students following the 1995 Oklahoma City bombing. *American Journal of Psychiatry*, *156*, 1069–1074.

Picou, J. S. (1996a). Compelled disclosure of scholarly research: Some comments on 'high stakes' litigation. *Law and Contemporary Problems*, *3*, 149–157.

Picou, J. S. (1996b). Toxins in the environment, damage to the community: Sociology and the toxic tort. In: P. Jenkins & S. Kroll-Smith (Eds), *Witnessing for Sociology: Sociologists in Court* (pp. 210–223). Westport, CT: Greenwood Press.

Picou, J. S. (2000). The talking circle as sociological practice: Cultural transformation of chronic disaster impacts. *Sociological Practice*, *2*, 77–97.

Picou, J. S., & Gill, D. A. (1996). The *Exxon Valdez* oil spill and chronic psychological stress. In: S. D. Rice, R. B. Spies, D. A. Wolfe & B. A. Wright (Eds), *Proceedings of the Exxon Valdez Oil Spill Symposium* (pp. 879–893). American Fisheries Society, Symposium, 18. Bethesda, MD.

Picou, J. S., & Gill, D. A. (2000). The *Exxon Valdez* disaster as localized environmental catastrophe: Dissimilarities to risk society theory. In: M. J. Cohen (Ed.), *Risk in the Modern World* (pp. 143–170). London: McMillan.

Picou, J. S., Johnson, G. D., & Gill, D. A. (2003). Transforming the "corrosive community": A participatory clinical intervention for chronic disaster impacts. Manuscript submitted for publication.

Picou, J. S., Marshall, B. K., & Gill, D. A. (2004). Disaster and litigation: Toward a model of chronic community impacts. *Social Forces*, forthcoming.

Picou, J. S., & Rosebrook, D. R. (1993). Technological accident, community class action litigation and scientific damage assessment: A case study of court-ordered research. *Sociological Spectrum*, *13*, 117–138.

Prezant, D. J., Weiden, M., Banauch, G. I., McGuinness, G., Rom, W. N., Aldrich, T. K., & Kelly, K. J. (2002). Cough and bronchial responsiveness in firefighters at the World Trade Center site. *New England Journal of Medicine*, *347*, 806–815.

Prince William Sound Regional Citizens' Advisory Council (1999). *Coping with technological disaster*. Anchorage, AK.

Purnick, J. (2001). Suing the city for Sept. 11? Oh, why not? *New York Times*, December 13, p. 1. Retrieved October 8, 2002, from LexisNexis™ Academic.

Quarantelli, E. L. (1991). Disaster response: Generic or agent specific. In: A. Kreimer & A. Munasinghe (Eds), *Managing Natural Disasters and the Environment*. Washington, DC: Environment Department, World Bank.

Quarantelli, E. L. (1992). The case for a generic rather than agent specific approach to disasters. *Disaster Management*, *2*, 191–196.

Quarantelli, E. L. (1998). *What is disaster? Perspectives on the question*. New York: Routledge.

Quarantelli, E. L., & Dynes, R. R. (1977). Response to social crisis and disaster. *Annual Review of Sociology*, *3*, 23–49.

Risen, J. (2002). White House drags its feet on testifying at 9/11 panel. *New York Times*, September 13, p. 12. Retrieved October 8, 2002, from LexisNexis™ Academic.

Robinson, W. I. (1996). *Promoting polyarchy: Globalization, U.S. intervention, and hegemony*. New York: Cambridge University Press.

Scelfo, J., & Smalley, S. (2002). The air down there. *Newsweek* [electronic version], September 10. Retrieved October 13, 2002, from http://www.newsweek.msnbc.com

Schuster, M. A., Stein, B. D., Jaycox, L. H., Collins, R. L., Marshall, G. N., Elliott, M. N., Zhou, A. J., Kanouse, D. E., Morrison, J. L., & Berry, S. H. (2001). A national survey of stress reactions after the September 11, 2001, terrorist attacks. *New England Journal of Medicine*, *345*, 1507–1512.

Sept. 11 survivors sue (2002). *New York Times*, August 16, p. 11. Retrieved October 8, 2002, from LexisNexis™ Academic.

Silver, R. C., Holman, E. A., McIntosh, D. N., Poulin, M., & Gil-Rivas, V. (2002). Nationwide longitudinal study of psychological responses to September 11. *JAMA*, *288*, 1235–1244.

Turkel, G. (2002). Sudden solidarity and the rush to normalization: Toward an alternative approach. *Sociological Focus*, *35*, 73–79.

Van Voris, B. (2002). Are toxic lawsuits in the air after 9/11 lawsuits? Federal fund may provide little help. *Inside*, *24*(24), A1. Retrieved October 8, 2002, from LexisNexis™ Academic.

Warily circling the Sept. 11 fund (2002). *New York Times*, June 5, p. 26. Retrieved October 8, 2002, from LexisNexis™ Academic.

Waugh, W. (1986). Integrating the policy models of terrorism and emergency management. *Policy Studies Review*, *6*, 287–300.

Webb, G. R. (2002). Sociology, disasters, and terrorism: Understanding threats of the new millennium. *Sociological Focus*, *35*, 87–95.

Weinbaum, E., & Page, M. (2002). Compensate all 9/11 families equally. *The Christian Science Monitor*, January 1. Retrieved October 4, 2002, from http://www.csmonitor.com/2002/0104/p11s1-coop.html

Weiser, B. (2002). Judge says Sept. 11 families can change minds on suing. *New York Times*, September 4, p. 3. Retrieved October 8, 2002, from LexisNexis™ Academic.

World Almanac and Book of Facts (2002). Facts About the September 11 Attack. World Almanac Education Group, Inc. Retrieved October 8, 2002, from LexisNexis™ Academic.

RECONSIDERING CONVERGENCE AND CONVERGER LEGITIMACY IN RESPONSE TO THE WORLD TRADE CENTER DISASTER

James M. Kendra and Tricia Wachtendorf

ABSTRACT

The World Trade Center disaster generated many of the features seen in other disasters in the U.S., including post-disaster convergence. We conceptualize emergency management activities as taking place within a multilocational "response milieu," and we suggest that the study of convergence should focus on the negotiated legitimacy of people in and wishing to enter it. We discuss the five types of personal convergers and how the access of each of these groups to the response milieu was related to their legitimation status. We then identify two additional forms of convergence: supporters or fans, and those who came to mourn or to memorialize. We conclude by discussing implications for policy.

INTRODUCTION

By virtually any definition, the World Trade Center attack on September 11, 2001 constituted a major disaster in terms of immediate loss of life and direct and indirect financial losses. Millions watched the event on television and observed the disaster

Terrorism and Disaster: New Threats, New Ideas
Research in Social Problems and Public Policy, Volume 11, 97–122

ISSN: 0196-1152/doi:10.1016/S0196-1152(03)11007-1

almost from its onset through the (admittedly selective) eye of media cameras. As might be expected given this level of exposure and the obvious magnitude of death and damage, volunteers as well as government officials, media, researchers, and others converged in great numbers on New York.

Convergence typically generates a tension since it both introduces needed resources and presents additional management challenges for public officials already occupied with their emergency duties. Researchers have long recognized convergence as a post-disaster phenomenon; it is, for example, discussed in what is widely viewed as the first sociological study of a disaster, Samuel H. Prince's investigation of the Halifax munitions ship explosion in 1917 (see Scanlon, 1992). Fritz and Mathewson (1957) presented perhaps the earliest comprehensive treatment of the subject. They defined convergence as "movement or inclination and approach toward a particular point" (1957, p. 3). They noted that convergence could be a "major hindrance to organized relief efforts," though later researchers have observed that volunteers can bring benefits as well as challenges (Stallings & Quarantelli, 1985; Tierney et al., 2001). Convergers may, for example, bring certain abilities that do not exist in sufficient quantities in the extant response organizations; they may already be close enough to damaged areas to provide immediate assistance; and they may provide for the flexibility that is needed when organizations confront rapidly-changing conditions.

Both the severity of the attacks on September 11 and their obvious national-security implications affected the distribution of the convergence activity and the motivations underlying it. Seldom in the U.S. does a disaster seem to be of national scope. Polling data indicated that some 86% of Americans saw the attack as an act of war, while 87% of Americans thought it was the "most tragic event in their lifetime" (Saad, 2001). The direct impacts reached beyond the Mid-Atlantic region. Airports across the country closed for days as a result of the terrorist threat and, although most of the casualties came from within the city and nearby communities, people across America and around the world lost family and friends in the New York City attacks. These impacts drove home the sentiment that there was a national if not international connection to the disaster and that a national, if not an international, response was necessary. As a result, volunteers arrived not just from elsewhere in the city, or New York State, or the very proximate Connecticut or New Jersey, but from much more distant areas. In addition, the high number of casualties among emergency response workers, rare in disasters, kindled a sense of professional camaraderie or affinity among emergency workers nationwide. Other features of the World Trade Center disaster were similarly conducive to massive convergence. These features included the suddenness of the disaster; the continuous media attention; and the number of casualties and their connection to many social groups at local, national, and international levels. The urban setting of the disaster, with its high local population density and (even during early phases of the emergency) availability of transportation, combined with the concentration of damage in a relatively small area, all set the stage for sizable convergence.

This paper focuses its attention on convergence of people, or personal convergence (The convergence of information and that of materials merit their own discussion. For more information on these convergence types, see Barton, 1969; Clifford, 1955; Fritz & Mathewson, 1957; Neal, 1992, 1994; NORC, 1954). The work of Fritz and Mathewson is a seminal work for the study of personal convergence. While the personal convergence categories exhibited in response to the World Trade Center disaster correspond with those offered by Fritz and Mathewson some fifty years ago, we contend that the authors' heuristic device that characterizes convergence with respect to a sequence of concentric circles is limited in its ability to adequately explain disaster convergence. The same criticism can be lodged against their notions of "movement toward the disaster-struck area from the outside – *external convergence* – and movement toward specific points within a given disaster-related area or zone – *internal convergence*" (Fritz & Mathewson, 1957, p. 3. Emphasis in original). Scanlon (1992) observed that people and supplies converge to various locations, and we contend that this complex of people and places constitutes a shifting and dynamic response milieu. We suggest that the study of convergence should focus on the legitimacy people claim for entry into the response milieu. To support this suggestion, we consider the five types of personal convergence offered by Fritz and Mathewson with respect to their negotiated legitimacy and the challenges associated with those negotiations. In addition to the five convergence types – the returnees, the anxious, the helpers, the curious, and the exploiters – we outline two additional types of convergers: the fans or supporters, and the mourners or memorializers. We further highlight the issue of legitimacy by showing how "helpers" were able to detect, or to create, opportunities to help. In some cases this involved establishing or borrowing their legitimacy; in other cases they operated at either a conceptual and/or physical boundary, satisfying their desire to help and providing services without having fully accredited or legitimized access.

Method

A field team from the Disaster Research Center (DRC) arrived in New York City on September 13th, 2001 – two days following the attack on the twin towers. Over a two-month period, DRC staff conducted over 750 collective hours of systematic observations and informal interviews at many sites significant in the emergency response, such as the city's Emergency Operations Center (EOC); the

Jacob Javits Center, a large convention facility where volunteers and donations were marshaled; the Family Assistance Center; command posts located close to Ground Zero; respite centers for emergency responders; and supply warehouses and staging areas. The data we used for this research primarily consist of hundreds of pages of documentary material collected in New York, the observations in the field, and the informal interviews. In addition, we draw upon articles from the *New York Times* and other newspapers that describe the types of people who converged to New York City and the various response locations following the attacks.

RECONSIDERING THE SPATIAL PATTERNS OF CONVERGENCE

Fritz and Mathewson appreciated the spatial dimension of convergence. They noted, for example, that because of population mobility in the United States, convergence could take place over long distances. Space and the spatially-referenced categories of internal and external convergence were significant in their discussions of convergence. Although their work is nearly fifty years old, *Convergence Behavior in Disasters* stands as one of the few comprehensive examinations of convergence in its various forms (Scanlon, 1992 provides the most recent detailed consideration of convergence activity, and there have been other works which focus on the management challenge). We therefore believe it is worthwhile to examine the models of convergence that Fritz and Mathewson presented. Although a useful heuristic, we find that their model of convergence patterns has certain limitations. Fritz and Mathewson characterized the spaces of convergence as follows:

> The center circle will be designated the "disaster area," the circle immediately surrounding it, the "contiguous zone," the third circle, the "proximate zone," and the final circle, the "remote zone." The contiguous zone should be thought of as a narrow band or strip immediately surrounding the site of destruction and casualties. The proximate zone includes the surrounding communities or satellite communities in the general environs of the disaster-struck community or area. The contiguous zone and the proximate zone together comprise the areas which are most likely to learn of the disaster immediately and from which the earliest outside response and assistance are likely to come. They are also the zones to which survivors go or are taken for immediate medical treatment, emergency relief, and shelter (Fritz & Mathewson, 1957, p. 5).

Fritz and Mathewson argued that the spatial pattern of convergence was a series of concentric circles, increasing in size: the disaster area, the contiguous zone, the proximate zone, and the remote zone. They described the contiguous and proximate zones as being the areas "which are most likely to learn of the disaster immediately and from which the earliest outside response and assistance are likely to come" (1957, p. 5). They then use this delineated space as a way of classifying convergers: the external convergers who cross zones toward the disaster site, and the internal convergers who operate within particular zones and who move toward relevant sites within a zone. The zones are thus also defined by how people within them participate in the response, and therefore these zones are the resultants of the interaction of social forces and physical phenomena.

Interestingly, there is little in behavioral responses outlined by the authors that distinguish between contiguous and proximate zones; the distance people move seems to be the key distinction. Fritz and Mathewson were less explicit about what constitutes the remote zone; it seems to be simply what is beyond the contiguous and proximate zones. The disaster zone is most explicitly a terrestrial or geographic feature; the contiguous zone is mostly based on geographic proximity but partly also based on the likelihood of assistance; even more than the contiguous zone, the proximate zone is shaped by the potential for offering aid; the remote zone is an indeterminate source of supplies or convergers.

It is difficult to discern to what extent these zones are defined by physical geography, and the extent they are defined by the movement of people either to or from them. The contiguous zone is defined as a "narrow band or strip," which gives a strong terrestrial underpinning; similarly, the proximate zone includes surrounding communities, which is also a strong terrestrial referent; yet both are explained as comprising the areas from which help comes and to which people go. To complicate matters still further, defining the "disaster area" is hardly straightforward. According to Fritz and Mathewson's spatial model, the size of the central disaster area obviously affects the scope of surrounding rings and therefore the classifications of convergence based upon those rings. But consider the September 11th, 2001, disaster in New York City. Would Fritz and Mathewson include only the former site of the World Trade Center complex as the disaster area? Or would they also include the areas that were pummeled with debris? What about the zone below Chambers Street where staging areas were eventually established, or the areas that remained evacuated for months, or the areas below Canal Street that remained evacuated for over a week, or those below 14th Street evacuated for days, or Manhattan, or New York City, or the surrounding communities where hundreds of those who died had resided? Without a clear, static definition of the disaster impact zone, it is difficult to consider how we can begin to draw clear lines of where the contiguous and proximate zones begin or end or if such a distinction is at all useful in understanding the convergence phenomenon.

To consider the point further, there would at first seem to be little controversy with respect to labeling the Ground Zero area (essentially the World Trade Center site) the disaster area in Fritz and Mathewson's model. But immediately the contiguous zone is more problematic. Along with the contiguous zone, the proximate zone is the area from which relief is likely to come, so this zone might comprise the area for several blocks around WTC, where people fled, got water, or ducked into shops, and from which others rushed in to help. An alternative view could be that this area, still in range of debris, is the disaster area, and would therefore push the boundary of the contiguous zone further away. Whether a converger is an "internal" or an "external" converger would shift with the definitional boundary of what constitutes the perimeter of the different zones. The disaster area could be seen as crystallizing at Canal Street, where police set up barricades, and south of which normal commerce, traffic, and domiciling ceased. This would nevertheless be a socially-constituted border. It would be a function of the topography of Manhattan and police officer decisions, as well as a conception of disaster that includes more than merely physical damage, since the buildings this far out were undamaged. It is also possible to widen the scale of analysis. The *New York Times* (August 21, 2002) published maps showing the spatial distribution of victims' residences. These cover many counties in New York, New Jersey, and Connecticut as well as other states (based on zip codes, the coverage does not extend internationally). The disaster area could thus extend more widely, as indeed it did in the federal disaster declaration.

Consensus on the scope, configuration, and relationship between these zones, in the World Trade Center attack or in any other disaster, is likely impossible. If we accept Scanlon's (1992) observation that people move to a variety of places other than the disaster area, then convergence is more generally a movement toward a point of significance identified as such by those who are converging, rather than a transgression of subjectively defined boundaries. We suggest that it is more helpful to view convergence as movement toward the *response milieu*, which we hold to be the complex of people and places involved in the response, a multilocational field of social activity. It is a shifting, protean entity which is created by participants and would-be participants acting across the variety of spaces and places involved in the response and helping to define those places as well. Just as the WTC attack expanded the imagination of people about disasters in the United States, it also expanded the scope of the people and places that could be converged upon. For it is not just places that are points of convergence – it is people, too. Convergers are themselves converged upon, without regard to the place or to any zone. Media converged upon then-Mayor Rudolph Giuliani as he was evacuating from the Trade Center complex in the early hours after the attack and continued to converge around him at press briefings. At the same time, family members converged on hospitals looking for missing loved ones while media, curious bystanders, and agencies attempting to provide assistance converged to where family members gathered. Convergers themselves create places to be converged to; such places do not always exist beforehand. The convergent points are thus a product of convergence activity and a function of response decisions, rather than an a priori focus. Nor are such places themselves fixed in space: the boats used in the evacuation of lower Manhattan became, in many instances, marshaling areas for other convergence.

It may be that any conception of zones surrounding an impact area is so variable with events that there is no real point in relying upon them as an analytical or descriptive tool. Scanlon (1992) makes little reference to zones, which suggests that he found them of limited usefulness as an analytical device. Scanlon examines convergence without reference to distinction between internal or external convergers. Even Fritz and Mathewson usually referred to the contiguous zone and the proximate zone with little distinction between them.

THE ACQUIRED AND CONTESTED LEGITIMACY OF CONVERGERS

The study of interactions at the boundary of the response milieu can be a fruitful line of research since the force of convergence is both felt at, and shapes, this boundary. People employ ad hoc strategies for claiming legitimacy while gatekeepers already within the response milieu come to accept or deny that legitimacy. Emergency managers often view convergence as a nuisance; convergers often view emergency managers as obstacles to their varied goals who restrict them illegitimately. Convergence, at its heart, presents a conflict of legitimacy: whose motives for converging are most acceptable or laudable, which if any convergers can provide the greatest good, and which convergers don't belong. We would therefore give primacy to the issue of legitimacy discussed by Fritz and Mathewson, arguing that legitimation transactions at the border of the response milieu provide considerable insight into understanding convergence as a phenomenon of collective behavior and perhaps the greatest insight into understanding convergence as an emergency management challenge.

Much research on disasters has sought to demonstrate the similarities in the aftermaths of various disasters, both as a means of facilitating the identification of generalizable principles for post-event response and as a means of enabling more holistic "all-hazards" planning. Disaster agents, however, retain important differences that are reflected in the disaster aftermath. Researchers now widely recognize the social construction of disasters: that maldistribution of resources, wealth, and power induces differential exposures and vulnerabilities. But disasters are not "constructed" only before an event, they also evolve as a construction after an event as well, through an ongoing social process of defining, reviewing, and retrospection. Convergence could thus be seen as a physical manifestation of the social construction of the disaster: simultaneously reflecting and helping to constitute it. Much of what we know of the World Trade Center disaster as a social event, incomplete as it is, is through watching where people went and what they did when they got there. The nature of the convergence, and the strategies and criteria for legitimacy, can be used as signals to detect the architecture of disaster as an evolving social and conceptual edifice.

We follow by examining seven types of convergers and discussing how their claims to legitimate access both shape the response milieu and at the same time provide challenges in securing and maintaining those contentious spaces. We should note that the categories of convergence are not mutually exclusive nor are they static. People may shift from one category to another as they acquire information or as the conditions change (Fritz & Mathewson, 1957). Yet each category of converger asserts different legitimation claims and offers its own challenges to emergency managers and other public officials. Moreover, each convergence type has different claims to accessing response locations and therefore plays different roles in shaping the boundaries to the response milieu.

Returnees

Returnees in New York City included residents, employees, business owners, and "substitute" returnees (Fritz & Mathewson, 1957) who are the relatives and friends of disaster victims who enter the area to assess the victims' losses and salvage their property. Owing in part to this country's high regard for property rights and other forms of locational tenure, returnees have a great deal of legitimacy when negotiating entry (Fritz & Mathewson, 1957). As the restricted zone surrounding the collapse site contracted – first at 14th Street, then to Canal Street, then progressively much closer to the World Trade Center – these returnees converged to the area, at first to obtain information, then to attempt to retrieve possessions, and finally to return home or to their places of employment. Returnees needed identification with a residential or business address to enter the secured area. Personnel guarding checkpoints were relatively generous when assessing people's identification; however, individuals were denied entry when they were unable to show an acceptable form of identification (driver's license, lease, phone bill). Returnees who had left identification documents behind during the sudden-onset evacuation and returnee substitutes were not given access. Employees who worked in area businesses were not allowed to enter secured zones even if they had left personal possessions behind during the evacuation. In the very early days after the Twin Towers collapsed, those charged with securing the area recognized the legitimacy of returnee claims to enter the most peripheral secured areas. Residents and business owners attempting to gain access to properties within only a few blocks of the collapse site were allowed in only under escort and for a brief period to retrieve a few important items such as money, documents, medication, and other easily portable possessions. The escorts ensured the protection of the returnees due to the extremely hazardous environment surrounding the residences and businesses closest to the area that became known as Ground Zero. We encountered returnees who waited over three hours for their group to be escorted, but nearly all accepted these restrictions without complaint and were polite when making inquiries of personnel who were staffing the checkpoints. Although the returnees were frustrated and tired of waiting, they could appreciate the dangers of the area surrounding the World Trade Center site as described by checkpoint personnel and documented by the television and print media.

While officials determined that returnees had legitimate claims to gain at least temporary entrée into the secured areas, the presence of the returnees posed particular challenges to the response. The NYPD, the National Guard, the New York Office of the Sheriff, and other agencies provided personnel to control the perimeters of the secured zones. These personnel needed to spend additional time and resources to determine the legitimacy of returnees. Security personnel and those who set the guidelines they followed had to weigh the needs of returnees against the number of escort resources, the safety of the returnees, and the continuity of the response effort. Personnel had to determine where returnees had legitimate claims to entrée and had to divide returnees into groups based on these divisions. Escorts also had to make sure that each member of the returnee group was accounted for within the zones and that the returnees indeed left the secured zone once they had retrieved their property.

Moreover, the World Trade Center site was considered a crime scene. The great distance that debris traveled following the towers' collapse raised the possibility that material pertinent to the criminal investigation could have found its way near or into the buildings to which returnees requested entry. Few would contest the legitimacy of returnees' attempts to retrieve important possessions from their disaster-impacted homes and businesses, particularly those whose homes were located in the secured and evacuated areas but did not suffer physical damage. This influx did, however, generate demands upon the city undergoing a protracted emergency response.

The Anxious

The anxious are people from outside the impacted area who attempt to obtain information about family and friends. In the case of this disaster, it is difficult to place a definite boundary to distinguish between those within and outside the impacted area.

Initially, many of the over 2800 people who perished were listed by the city as "missing." Relatives, coworkers, and friends came to various sites around New York City in search of the missing. Notably, the immediate convergence of the anxious did not only consist of those who lived in the impacted zone or even in Manhattan. The World Trade Center complex was a commercial space and many of those who worked in the Twin Towers and surrounding buildings commuted from elsewhere in New York City, New York State, New Jersey, Connecticut, and Pennsylvania. These anxious were the first to arrive in search of the missing. Closure of U.S. airports delayed the arrival of anxious convergers coming from great distances; however, thousands soon arrived after airports were reopened.

While many of the anxious gravitated toward Ground Zero, they more often converged toward other locations throughout the city that they believed would provide the most information regarding their missing loved ones. The convergers posted missing-persons notices in major tourist locations such as Times Square and on bus shelters and traffic light posts around the city. The highest concentration of anxious convergence could be traced to areas with the highest concentration of missing person notices. Hospitals such as St. Vincent's, New York University, and Bellevue were in close proximity to Lower Manhattan and therefore sites where anxious convergers were hoping to find loved ones who might be too injured or traumatized to identify themselves. For many people the anxious period was more protracted than we have seen in other more recent American disasters, since many remained in an anxious state until remains were found and identified.

City officials recognized the legitimacy of the anxious convergers and sympathized with their plight. At the same time, officials faced the challenge of meeting the needs of thousands of family members and friends looking for loved ones. They established sites where they could meet those needs; they directed the anxious to those locations and provided them with credentials. For example, at the Pier 94 family assistance center, family members of victims and the missing received many services such as information regarding the rescue and recovery process, a means of registering the missing, a depository for DNA identification samples, free international telephone and email access, counseling, food, child care, assistance with paperwork, and the opportunity to connect with other families who had also suffered direct personal loss as a result of the disaster. To better meet the needs of the anxious, the family assistance center changed locations several times to more appropriate facilities. City personnel also adapted procedures to eventually ensure the privacy of the family members, guarantee their safety from any potential threats, provide space and access to supporting agencies, provide a comfortable and compassionate environment, and allow support and response workers to complete tasks assigned to them.

At first, security at sites such as the Lexington Street Armory was somewhat porous. Although security was present, it was not difficult to gain access into this family assistance center. Eventually, as time passed and as the family assistance center moved to Pier 94, the anxious close associates of the missing were distinguished from: (1) those who had a less direct connection to the missing persons – such as coworkers or friends; and (2) the generally anxious – those who had not lost a family member but were worried about the general safety of loved ones due to a perception that New York City was under a persistent yet ambiguous threat of future attacks. The anxious close-associates were viewed by the city and by the general public as a more legitimate contingent than the generally anxious; they received credentials and, in general, more sympathetic and compassionate treatment. The generally anxious convergers could locate counseling services through telephone numbers advertised throughout the city; however, additional support and services were not provided to this latter group. In contrast, anxious close-associates of the missing were given identification badges that allowed them access to the assistance centers and helped security and support personnel distinguish them from other anxious convergers as well as from the general public. The credentialing process was established to ease the entry of anxious close-associates of the missing into key facilities. It also signified to support workers their status as close-associates of victims who should be treated with patience and compassion.

Helpers

Helpers converge in order to assist victims or responders. The category of helper has proved to be particularly troublesome in disaster response, because helpers seek not just to enter the response milieu but to take on roles that will have a constitutive effect as well. Helpers present their own legitimacy claims, but these can have an oppositional effect, sometimes challenging the adequacy of the response. A tremendous number of helpers came to offer assistance in lower Manhattan and the support stations established throughout New York City. Some helpers were already part of the local response milieu, such as emergency response personnel; others were from outside the local jurisdiction but came under the auspices of existing mutual aid agreements from federal, state, or other local jurisdictions or as part of expanding organizations – organizations with new structures but whose normal tasks included disaster work (Dynes, 1970). Many ad hoc volunteers were associated with extending groups (that is, groups with static organizational structures but new to disaster tasks). The rest were individuals eager to help in any way they could.

Convergence and emergence, another important disaster phenomenon, are in many ways complementary. Stallings and Quarantelli (1985, p. 94) define emergent groups as "private citizens who work together in pursuit of collective goals relevant to actual or potential disasters but whose organization has not yet become institutionalized." Converging volunteers often gravitate to groups that have emerged in response to disasters, either to provide additional support or to perform tasks that complement those of emergent groups. At the same time, by providing ever-larger numbers of volunteers, convergence sets the stage for further emergence or for new groups to develop. Public officials often find emergence especially onerous during the response phase of a disaster, since the appearance of these groups can suggest the inadequacy of official response efforts. "Municipal law enforcement in particular, [tends] to view volunteer emergence as a crowd control problem" (Kartez & Kelley, 1988, p. 140). The nature of the World Trade Center attack intensified this concern.

Volunteers were conspicuous at all locations of emergency response activities. It is important to clarify, though, that not all volunteers were civilians: police and firefighters from nearby localities, adjacent states, and (later) from across the country dispatched themselves to New York or were requested by the City of New York. Many volunteers had demonstrably relevant skills: for example, physicians and nurses, counselors, and other health-care workers arrived, as did workers from the construction trades. Not all volunteers were without formal affiliation. Many worked with the American Red Cross, perhaps "the best known example of institutional volunteering" (Tierney et al., 2001, p. 112), the Salvation Army, and other expanding (Dynes, 1970) disaster agencies with a long history of incorporating and organizing volunteers. Other volunteers were coordinated and deployed by their trade union organizations.[1]

Though helpers of every description arrived, many volunteers were without immediately applicable skills, training, or connection to the recognized emergency management apparatus. They wanted to help, too, but their ties to the response milieu were rendered problematic by their lack of identifiable, relevant capabilities (or capabilities not needed at the time of their arrival), lack of legitimacy or connection to an organization from which they could borrow legitimacy (some officials worried about liability issues), and probable lack of familiarity (from the perspective of public officials) with emergency operations. Legitimacy has a strong geographic dimension: there is also evidence that "not being from New York" discounted to some extent the qualifications of some outside emergency professionals. Not only did these outsiders appear to lack local knowledge, but local emergency workers claimed the right to control access to the site because they had family and friends among the victims.

Helpers present particular challenges because their desire to help is often not matched by their ability to be integrated into the response milieu. One of the primary requirements for an effective response to disasters and other complex crises is the ability of all the participants in the response to develop, maintain, and act upon a "shared vision" of emergency needs, goals, and available resources (see Comfort, 1999). Typically, within a single organization, the shared vision is developed both explicitly – through training, drills, and exercises – and implicitly – through steady immersion in operational activities so that the organization's norms are imparted by exposure to them. The process is, of course, an imperfect one; dissident groups and resisters exist in any organization, and socially-negotiated subcultures exist alongside those preferred by corporate officers. While no organization functions according to the formally-delineated structure, sufficient congruence exists by consent or through enforcement to ensure that organizational goals are attained. Loss of shared vision constitutes a management crisis that demands intervention (see, for example, Brenneman, 2000).

Volunteers are outsiders, yet they are outsiders with a halo of political significance, especially because of their patriotic motivation. Put bluntly, emergency managers recognized that outright rebuffs of volunteers constitute bad public relations. They saw that the volunteers had some kind of legitimacy, even if they themselves were skeptical of the basis of that legitimacy. Some volunteers were offended that they were not allowed to assist. Yet at the same time, officials were not turning away volunteers out of disregard for the outpouring of generosity, or because they felt that the appearance of volunteers indicated some sort of failure of emergency planning. Rather, they appreciated the intentions of the volunteers and the community spirit that represented. Nevertheless, because the disaster was caused by a terrorist attack, there was great concern for identification and credentialing of all personnel working at the various locations. Within this context, volunteer personnel became another group that needed to be accounted for, and therefore potentially a distraction that outweighed their utility. Some emergent jargon developed that reflected a burgeoning oppositional relationship between emergency managers and donors. For example, the terms "rebel food" and "renegade supplies" were used to refer to food and other items that were brought into the impact area on an ad hoc, uncontrolled basis, frequently by people who were unknown to the recipient. Food items ranged from home-baked goods to hot meals, typically pasta. Soon after the attack, a well-known steakhouse chain (one of many food-vendors who converged) set up a barbecue near the disaster site, an act that offended some of the firefighters. In particular, an Urban Search and Rescue official bitterly condemned the barbecue at a staff meeting, calling the "back-yard barbecue" atmosphere disrespectful to the gravity of the situation and to the victims. In contrast, some rescue workers appreciated the variation in diet the barbecue introduced, and found that this form of volunteerism made positive contributions to the response. Not only did those within the response milieu need to contend with the influx of people to the area and with the health and liability concerns associated with the distribution of food, but they also had to balance public relations and appreciation of assistance with the added complexity these volunteers posed for security.

Every skill that might be required in an emergency response was available in New York City, but determining the professional qualifications of volunteers was difficult. After September 11, not only was the temporal extent of the risk uncertain (would another attack follow?), but the subsequent anthrax episodes heightened the perception of being under constant threat. Over time, as the response evolved, volunteers were less and less welcome; it may even be possible to mark the shift from response to recovery by noting when non-affiliated volunteers were asked to leave or were evicted from various sites. A week after the event, some officials stated that they didn't "want volunteers in even if they're already here" and they were citing as a problem "continual freelance medical practice," referring to unauthorized people treating the injured and providing counseling.

A further complication was the persistence of considerable hazard even after the acute stage of the disaster – the attack and collapse of the towers – was over. The site remained dangerous: broken glass and debris, smoke and gases billowing out for months, structural steel precariously embedded in surrounding buildings. The kind of labor-intensive work that can be performed by relatively untrained volunteers, such as sandbagging or clearing generally innocuous debris, was largely impossible at Ground Zero. In addition, the relatively concentrated space of the Ground Zero area (much more concentrated than in hurricane, flooding, or even tornado events) limited the number of people who could safely work there.

Not all processes of convergence are equally problematic. The waterborne evacuation of Lower Manhattan, in which the Coast Guard estimates that more than 500,000 people were evacuated by boat, is an instance of convergence in which potentially troublesome aspects (the security risk created by so many boats near Manhattan) were subsumed by the need to move people away. Other volunteers integrated into the emergency response were GIS specialists from local colleges who were instrumental in producing maps and images based on remote-sensing data. A military reserve officer volunteered his services (and was eventually issued official orders to do so) and became an invaluable assistant to a key response official. For example, the officer took notes at planning meetings, coordinated transportation for operations staff, and spent time finding solutions for various operational exigencies. What is especially significant about the role of such integrated volunteers is that they had a particular skill or set of capabilities or attributes that were congruent with the ambient response needs, and that were simultaneously obvious to the emergency managers who were in a position to "hire" them. For example, several chiropractors talked their way into the Emergency Operations Center and performed massage and spinal adjustment. They allied themselves with the Red Cross initially in order to maintain access to the facility, but they were able to offer a specific skill, which, even if not planned for in advance, seemed useful to those who already had access. A police officer from Illinois stood guard duty: again, this was a task he could perform with little supervision and which enhanced the existing personnel resources.

The Curious

Curious convergers come to the impacted site primarily to view the destruction left in the wake of the disaster and the activities surrounding the response. These include people with minimal personal connection to the disaster victims. Initially, this behavior is exhibited by "persons in the disaster area who are not preoccupied with pressing personal concerns. Later, however, the initial curiosity seekers are likely to be augmented by survivors who were formerly anxious about their own or other's safety and welfare, and by persons who enter the disaster area from outside" (Fritz & Mathewson, 1957, p. 47).

In New York City, some people gathered to witness the spectacle of the disaster while others simply tried to get as close to Ground Zero as they could. In the words of one high-placed public official, they wanted to "touch the dust." Onlookers gathered at checkpoints and barricades, hoping to catch a glimpse of what remained of the Twin Towers, moving closer to the World Trade Center as the secured perimeter contracted. Onlookers generally adopted a reverent demeanor and seemed awestruck at the damage, even though only a portion of it was visible from the perimeter. Some became quite determined in their efforts to catch a glimpse. Many onlookers took photographs while a few even climbed scaffolding along neighboring buildings in order to get the best view. The density of the crowds was at times so thick throughout the day that it became difficult to pass along the sidewalk. The onlookers continued to converge long after the response and early recovery periods. The presence of some disaster tourists was accompanied by potential financial or political dividends for a city in recovery.

Curious convergers were not without their challenges. Because the crowds stood at the edge of the perimeter rather than inside the secured area, they could not be legally prohibited from taking photographs. Still, their presence created a tension for security personnel guarding the area. Although NYPD could not prohibit the public from stopping and looking towards Ground Zero, police encouraged (often with limited success) crowds to move along to allow sidewalk traffic to proceed.

Some responders and people who lost loved ones could not understand the curiosity of the convergers and were noticeably shaken at the sight of so many onlookers.

For over two months, police officers forbade photography as far as two or three blocks from the perimeter, which itself was two blocks away from the disaster site, saying that it was a "crime scene." When asked for further details, they said only that they had orders to restrict photography. Other authorities claimed that they wanted to maintain the privacy of those working at the site, including personnel searching for remains that were often those of colleagues, friends, and family members. In contrast, other guidelines stated that photography was forbidden within the Ground Zero barricades, but merely discouraged or censured beyond the primary secured area. What is significant with respect to our study of convergence is the varied status of the site and the justifications that status allowed in controlling or restricting people's movements or actions. Notably, some officers appeared to understand the desire of the crowd to see the damage, demonstrated patience while encouraging the flow of traffic, and tolerated photographs of the physical destruction. Regardless of the attitude of personnel toward the general public, they had to consider public safety, maintain the flow of foot and vehicle traffic, and ensure that crowd members did not attempt to gain unauthorized entry into the secured zone. The closer the perimeter to Ground Zero, the more difficult this task became.

Members of the general public were not the only ones to gather near the collapse site to witness the destruction. From the very beginning of the disaster event, media converged to New York City in search of the ever-developing story. The media is very important to an administration in a disaster-impacted community for it is through this outlet that governing agencies provide information to the public. But officials must also work to ensure that converging journalists are safe, that correct information is being distributed, and, in the case of this disaster, that family members and emergency response workers are not disturbed nor security measures compromised. Although the press briefing room of the Mayor's Office of Emergency Management was destroyed with the loss of the Emergency Operations Center at 7 World Trade Center, the City was relatively quick in re-establishing a press briefing area at the Police Academy later in the afternoon of September 11th. Other agencies assisted in issuing media releases that directed the media to relevant briefing locations. Still, some media were ill-informed of press marshaling areas and many of the personnel staffing various checkpoints could not direct the media to a central information center, perhaps because of the destruction of the EOC, perhaps because marshaling points were continuously changing in those first few days, perhaps because of heightened security.

Security personnel sometimes directed the media to cordoned-off areas as a means of clearing space and limiting the proximity of the press to response efforts.

Yet, unlike the public, who were often restricted from photographing (even from a considerable distance) the rescue and recovery efforts, the relevant facilities, missing person notices at family assistance centers, or the World Trade Center site, the media with their high-powered zoom lenses were able to videotape and photograph these same scenes and broadcast them to the public. In a sense, media physically converged to response-related sites but were able to optically converge closer to the sites and closer to those impacted by and responding to the disaster because their camera lenses have such a far reach. For example, while officials were somewhat able to maintain the distance of photographers from family members and response workers viewing the missing person notices at the family assistance centers by directing them toward a designated press area, television cameras were able to zoom in to a tight close-up of the individuals.

Dignitaries including the President of the United States and heads of state from other countries also toured the site to witness the destruction firsthand. Inevitably, additional resources and precautions were needed as these dignitaries visited the site, but at the same time, their desire to see Ground Zero was accommodated as local officials believed that actually seeing the enormity of the destruction was instrumental to efforts to garner political and financial support. Nevertheless, some officials regarded the visiting dignitaries and celebrities as a distraction. The coordination of their arrival was a logistics problem requiring vehicles, additional safety precautions, and escort personnel. While the presence of dignitaries was viewed as legitimate, officials preferred that no celebrities visit. Aware that such visits could not be stopped, an official asked in one planning meeting, "Can we limit them?" In the end, it was difficult for officials to control the entry of movie stars, television personalities, and sports figures, who could persuade some security personnel to allow them special access.

The mass influx of the general population near the restricted site generated challenges associated with crowd-control, health and safety, security, and the emotional impact of the crowds on site-workers who had lost family and friends in the attack. At the same time, the curious asserted their legitimacy by showing they were often motivated by a sense of reverence for the site and did not intend to obstruct or complicate the disaster response. They merely wanted to see for themselves the destruction that had resulted from the attack.

Exploiters

Exploiters are convergers who use the disaster for personal gain or profit. In the days and weeks after the attacks, reports emerged of malicious websites falsely stating that some individuals who were listed as missing had indeed been found.

While little financial profit could be gained from these websites containing false information, the pain and anxiousness of family members were exploited, presumably for the amusement of those publishing the information.

Officials were concerned about the potential for financial exploitation and identity theft posed by the missing-persons notices posted throughout New York City. Family members had included on these notices such detailed information as social security numbers, passport numbers, and dates of birth. Security and other staff at the family assistance centers attempted to maintain a delicate balance between sensitivity to family members' grief and protection against fraud and exploitation. They did not prohibit family members from displaying this confidential information; instead, officials prohibited passer-bys from photographing or recording any information from the missing person notices posted at the family assistance centers. Still, these notices were posted around the city and staff could not monitor every lamp-post or bus shelter. Furthermore, concern about potential identity theft was not shared by family members, who believed that wide dissemination of the information on the notices would lead to the recovery of their loved ones. It is difficult to determine the extent to which identity theft actually took place, despite the concern by security personnel and the media. Some statistics are available on the number of individuals charged with other fraudulent activities related to the World Trade Center attack. A special World Trade Center task force was developed – called "Vulture Sweep" – to investigate cases of fraud. According to the Office of the Inspector General, as of April 15th, 2002 twenty-three people had been charged with falsely filing for death certificates, claiming that members of their family died in the September 11th attacks on the World Trade Center (Eye on OIG, 2002).

Exploiters found other means to profit from this disaster. Vendors converged to the street and sold merchandise commemorating the World Trade Center or the September 11th attack. The vendors claimed portions of the proceeds were going to funds for victims, when in fact no donations were made or an extremely small percentage were donated. The extent of this fraud is difficult to track as most vendors did not explicitly state the percentage they would donate. Some may have been quite generous, while others may have kept all or most of the profits. The challenge for the purchasing and donating public was to distinguish the formal sources and informal yet well-meaning sources from illegitimate sources.

NYC departments tried to educate the public regarding bogus charities and, through various media outlets, encouraged the public to donate to any number of a set of recognized charities or fundraising efforts that the city deemed legitimate. Some firehouses even began selling FDNY and 9/11 t-shirts and pins in order to counter vendors who were making profits from their department's logos. Proceeds from the firehouse's sales were donated to the families of lost firefighters.

Although the above examples seem exploitive in nature, there were those who profited from the disaster but with arguably less malicious intentions. As curious "disaster tourists" converged to various locations, disaster entrepreneurs converged to curious convergers in order to profit from them. Photographs of the Twin Towers and T-shirts with patriotic or memorial themes were widely available in Times Square within 48 hours of the attack, and many such items bore the signs of hasty manufacture, such as misspellings or the wrong dates for the construction of the World Trade Center. Disaster entrepreneurs followed the secured perimeter as it contracted. The exploiters described above were clearly entrepreneurial in their actions, but not all disaster entrepreneurs were truly exploitive in nature. Many were identifying a market that was indeed an extension of a tourist market they routinely served. For example, vendors who typically sold prints of the Empire State Building, the Flatiron Building, the Chrysler Building, the Statue of Liberty, and the Twin Towers instead focused their sales on pictures of the Twin Towers. T-shirt vendors who sold inexpensive "I Love New York" products increased their stocks and added other shirts that essentially comprised a new memorial line of products. Vendors who had previously sold wallets, purses, and knick-knacks now added small American flags and pins to their available wares. Food and drink vendors took note of the crowds and recognized the potential for sales. Some of these vendors had not typically worked in the Lower Manhattan area while others returned to the vicinity of the World Trade Center where they had sold similar products before the disaster. They adapted the types of goods for sale to meet the demands of the convergers. Again, some found this practice inappropriate considering the degree of damage and large numbers of casualties in the disaster, yet many others did purchase the products. In addition, NYPD had the added burden of distinguishing legitimate vendors from the exploiters.

The Supporters

Two additional forms of convergence, ones not previously identified by Fritz and Mathewson, were evident in this disaster. The first of these new categories – what constitutes the sixth form of convergence – consists of fans or supporters. These convergers are either individuals or groups who gathered to encourage and express gratitude to emergency workers.

Notably, the convergence of fans and supporters was closely related to the sense of national crisis and the high number of rescue-worker fatalities. Groups of well-wishers gathered at intersections, particularly along the West Side Highway, waving flags, showing banners, and cheering as emergency service vehicles drove by. Unlike curious convergers, members of the supporter group were not motivated primarily by curiosity to witness the disaster's destruction, but rather wanted to demonstrate thanks and moral support to those impacted by or responding to the disaster. One newspaper article included the following account of those who converged to Ground Zero on Christmas:

> Among the spectators gathered along the perimeter of the site, Veronique Fallous said she had come from Paris to celebrate the holidays in New York.
>
> "I wanted to be here with a lot of people to feel part of the community", Fallous said.
>
> Miriam Neeson of Atlanta spoke through tears. She said being near the site was "a form of sharing. This is the only way we can give support."
>
> (Associated Press, 2001)

Some well-wishers converged to the same points as the curious, leaving thank-you notes and banners along fences as close as they could get to Ground Zero. Many, however, converged to staging areas and facilities at some distance from Ground Zero out of recognition that workers there also deserved appreciation. As the weeks went on, the thank-you notes and banners began to acknowledge the contribution of agencies such as the Mayor's Office of Emergency Management, Emergency Medical Service personnel, ironworkers, and construction workers, in addition to the Fire Department of New York City, the New York City Police Department, and the search and rescue dogs to whom many thank you notes were initially sent.

Supporters also sent and distributed food. Just as a neighbor may deliver a meal or baked goods to a neighbor who is ill or welcome a newcomer to the neighborhood, these well-wishers handed out food products as explicit gestures of support. Emergency responders appreciated these gestures, but some of them recognized the potential security risk associated with these unknown products. Could there be biological warfare agents, or poison, in the food? One public official expressed concern about a basket full of carefully zip-locked bags of chocolate chip cookies with notes from children thanking those involved in the response. Although he appreciated the donation and the accompanying sentiment, he questioned how one could be sure that the cookies were not, in fact, laced with a dangerous substance. Similarly, the driver of a sheriff's vehicle used for escorting officials to meetings near Ground Zero declined donuts being passed out by well-wishers. The driver was very pleasant to the supporters, stating that he had just eaten breakfast and thanking them for their encouragement. Afterward, he remarked that eating food given out on the street by passers-by was not normally prudent, and particularly not so after September 11th. If such concerns appear to be overreactions, consider that shortly after these incidents took place, Senator Daschle received an anthrax-laced envelope, listing a fourth grade class from a

New Jersey school as its return address. The security environment was one that legitimated their concerns, even though they knew that the vast majority of those distributing the food products were genuinely good-hearted.

Officials were aware that the distribution of food by almost all volunteers and well-wishers came with the best of intentions and thus acknowledged their legitimacy; however, these same officials understood they needed to balance appropriate public relations and appreciation with the persistent security threat. Despite the need for caution, the outpouring of thanks visibly touched many response workers. The same public official who expressed concern about the bags of cookies was brought to tears when recalling the many people who stood all day at the West Side Highway just to say "thank-you" again and again.

The Mourners and Memorializers

We identified a second convergence category beyond those described by Fritz and Mathewson: the mourners or memorializers. These were people who went to locations such as firehouses, Union Square, and Ground Zero to lay flowers, light candles, create memorials, and mourn the dead.

The supporters and the mourners/memorializers share some common characteristics. The same individuals may shift in and out of the categories or stand simultaneously in both. Nevertheless, there seems to be enough differentiation to warrant the consideration of a separate category, based on the actions of the people we observed and the texts they employed. Both supporters and memorializers used banners, but while the supporters' signs expressed appreciation for the response and recovery workers, the memorializers displayed banners paying homage to those who died, the impact upon the nation, and the destruction of the buildings themselves. The waving of a banner with a message of gratitude or a patriotic theme expresses a different sentiment from the laying of flowers or lighting of candles, which tend to be more associated with expressions of condolence, funerals, or eulogies. Admittedly there is a fuzzy boundary between the category of supporters and that of mourners/memorializers, but we sense that the defining attributes of these categories have distinct centers of gravity, a recognizable core even if the edges are somewhat nebulous.

An important consideration in the World Trade Center response is that the response itself was so protracted. The rescue and the recovery phases were not well-defined. Mourning and memorializing occurred simultaneously with both rescue and recovery efforts, posing further challenges for public officials especially with respect to site security, legitimate access, and the personal safety of those who wanted close access to the site. For example, just over a month after the attacks, a memorial service was held for family members at Ground Zero. Significant work went into preparing the area to accommodate the large influx of people into a persistently hazardous area. Despite the time invested, most of those we talked with who were involved with the recovery operations appreciated the desire for family members to participate in a memorial service at the site. FDNY personnel acceded to requests of family members for pictures and even offered very small pieces of debris for them to take home.

Memorializing, and mourning in particular, are actions with a high degree of legitimacy: some of the shrines, though established in an ad hoc manner, took on features of semi-permanence and were cared for and maintained, as at a firehouse we saw. Family members gained more legitimacy as they changed status from the anxious to the mourners. Efforts to accommodate family-member mourners extended beyond the memorial service. City officials built access ramps and organized escorted visits to overlook the site.

We do not argue that demonstrations of support, or of mourning or memorializing, are unique to the World Trade Center disaster. Flags, banners, and yellow ribbons are conspicuous in times of national crisis, and shrines of various levels of sophistication are evident after many disasters, particularly transportation disasters, as well as at the deaths of celebrities (see, for example, Eyre (1999) for a discussion of mourners and spontaneous expressions of grief after both disasters and tragic deaths such as the death of Princess Diana). But the actions of those who came to support, offer sympathy, mourn, and memorialize as we discuss here converged in the midst of the disaster response rather than following it. In most American disasters, the memorializers and mourners arrive after the response workers have moved on to a recovery phase. The challenges posed by mourners and memorializers in this disaster were heightened due to the protracted emergency response (where convergers were exerting legitimacy claims to the response milieu), the dangerous conditions, and security concerns.

CONCLUSION

Disaster planning must anticipate the proliferation of volunteers and other convergers. The ideal outcome for handling volunteers, as suggested in the literature (see Forrest, 1978; Stallings & Quarantelli, 1985), is that they somehow be integrated into the formal disaster planning process: that convergence and group emergence be converted from a seemingly unexpected phenomenon into an anticipated consequence that can be utilized. But where volunteers should fit in the burgeoning response and how all involved in that response can develop and maintain a shared vision of needs and capabilities remain difficult questions. The ad hoc volunteers, as opposed to volunteers associated with formal organizations such as the Red Cross, can avoid bureaucratic obstacles that delay the delivery of services, and they can fill in gaps left by organizations already active in the response milieu. At the same time, many of these volunteers generally lack the training and experience required to participate in the more complex aspects of emergency response and the arrival of large numbers of lay people may not be synchronized with the manifestation of needs that they can fulfill. The concern about ongoing terrorist activity would make the presence of volunteers even more challenging since they would need to be monitored more closely than usual.

The most "successful" helpers – those who established legitimacy, negotiated access with gatekeepers, and entered the response milieu – were those who were able to work with minimal supervision by official emergency workers. Important to establishing legitimacy is gaining access early in the disaster response, especially by establishing a legitimized relationship with an insider and negotiating changes in the disaster environment (especially security regulations). From a practical perspective, it may be possible to look at the attributes of the "successful" convergers and use those attributes as a basis for anticipating the kinds of activities that helpers can most readily be engaged in and then to provide those opportunities. Given that an important qualification of the successful volunteers was the capacity to work with little supervision at tasks that required little organizational integration, emergency management officials might anticipate the tasks that meet that description, and select volunteers accordingly. It may also be possible to leverage the enthusiasm of the volunteers and to assign them other useful work in other spheres of activity that might free up local personnel to work in the emergency response.

We can also extend practical ramifications from a shift in focus to legitimacy transactions at the border of the response milieu. For one thing, as convergers establish their legitimacy, the response milieu expands as they enter into it. These convergers can then become the focus of other convergers and can sometimes even perform gatekeeper roles: sometimes they are "deputized" into a quasi-official status. It also suggests that emergency managers, who usually prefer to keep out volunteers and to limit other kinds of convergence, would do well to focus on the legitimacy claims that emerge or that may emerge in a disaster. Fundamentally, the convergence "problem" is one of competing legitimacies, manifest as conflict over control of, and access to, space. Fritz and Mathewson observed that restricting access of returnees – the most "legitimate" of the convergers – requires justification. That legitimacy is founded on bedrock principles of locational tenure, but the observation really extends to the full spectrum of convergence activity. To help, to show support to those on the front line, to mourn, even to be curious, are themselves, if not bedrock principles, then at least foundational activities that are heralded in other spheres of life: they will not be suddenly abandoned in disaster nor is it certain that they should be. Emergency managers should consider, soon after a disaster of any type, how the nature of the event will give rise to patterns of convergence and be prepared to engage, and to speak directly to, the probable motives and legitimacy claims of convergers. Simply declaring a place to be a "crime scene" is not likely to be accepted as a sufficient rationale by convergers. Guards and barricades are always available to impede convergers, but that hardly seems an elegant response to a phenomenon with such powerful social antecedents. Rather we suggest that it may be possible to address the motives and legitimacy claims rather promptly, and to reduce the inclination to converge or to ease tensions when people do converge. For example, the police officers posted at checkpoints who were letting residents in for short periods were very effective at communicating with the returnees. As a consequence, the people were quite patient and orderly. It is true that the safety concerns that the officers articulated were fairly unambiguous, the evidence for which was quite obvious, and that eased the communications burden in this instance. Nevertheless, a careful and sensible elaboration of the issues can perhaps help in other situations, avoiding inconsistencies and absurd juxtapositions. During a news broadcast, the anchor reported that donations of goods should stop and that blood was no longer needed. But at the bottom of the screen, the news ticker gave instructions for sending donations and giving blood. Such an inconsistency is not the fault of emergency managers, but they need to be certain that messages to the media are clear and well-reasoned to minimize confusion and the dissemination of contradictory information. We noted earlier that people were forbidden to take photographs of the site from a distance of a few blocks because it was a crime scene. Yet, just one block further away, TV cameras were set up. The powerful lenses on these cameras easily produce detailed images despite the distance, certainly far more effectively than the disposable cameras carried by some passers-by.

Public officials must be prepared for the various manifestations of convergence and develop, if not detailed plans, then at least guidelines for handling it. Public officials have difficulty with convergers for reasons other than technical crowd-control or safety reasons. The usual public-private divide on which public policy is based is destabilized. "Bureaucracies are structured on an everyday basis to do things *to* or *for* people rather than with them, and this carries over into disaster situations" (Quarantelli, 1982, p. 7). Helpers, for example, take on the "office of citizen" with renewed vigor. Returnees invoke the legitimacy of their publicly-registered, publicly-sanctioned property rights. Some thought must be given, and provision made if possible, for accommodating the particular needs of returnees, the anxious, and the mourners and memorializers, all of whom are engaged in activities that are elemental in the face of crisis and loss.

Unquestionably, attending to convergers is challenging for officials, but it is a fact of life in emergencies. The choice is to keep convergers away by various kinds of force, or to understand the phenomenon of convergence and to try to affect its genesis and morphology. There are good and bad reasons for restricting the movement of people over public spaces and to various kinds of facilities. A sound public policy will consider convergence with respect to the multiple legitimacy claims that arise after disaster and with respect to emergency management necessities that can be clearly and coherently stated, and that will not create credibility-damaging contradictions.

NOTE

1. In our discussion of helper convergence, we have primarily focused on volunteer helpers. Clearly, helper convergence also stemmed from internal formal or official sources. Hundreds of city, state, and federal departments responded to the World Trade Center disaster in a non-volunteer capacity. Coordinated efforts also came from non-profit groups and the private sector. The formal organizational response was impressive and contributed to city's resiliency. At the same time, the coordination of non-volunteer helpers was a considerable challenge for New York City.

ACKNOWLEDGMENTS

This research is supported by grants to the University of Delaware Disaster Research Center from the National Science Foundation (supplemental to grant EEC-9701471), the Multidisciplinary Center for Earthquake Engineering Research, and the Public Entity Risk Institute (2001-70). The authors are responsible for the views presented here. An earlier version of this paper was presented at the 97th Annual Meeting of the American Sociological Association, Chicago, Illinois. August 16–19, 2002. We are grateful to Kathleen Tierney and the anonymous reviewers for their comments on earlier drafts of this paper.

REFERENCES

Associated Press (2001). Search effort at Ground Zero doesn't stop for the holiday. *Atlanta Journal Constitution*, December 26, A13.

Barton, A. H. (1969). *Communities in disaster: A sociological analysis of collective stress situations.* Garden City, NY: Doubleday.

Brenneman, G. (2000). Right away and all at once: How we saved Continental. In: *Harvard Business Review on Crisis Management.* Boston, MA: Harvard Business School Press.

Clifford, R. A. (1955). *Informal group actions in the Rio Grande flood*. A report to the Committee on Disaster Studies, National Research Council.

Comfort, L. (1999). *Shared risk: Complex systems in seismic response*. New York, NY: Pergamon.

Dynes, R. R. (1970). *Organized behavior in disaster*. Lexington, MA: D.C. Heath.

Eyre, A. (1999). In remembrance: Post-disaster rituals and symbols. *Australian Journal of Emergency Management*, *14*, 23–29.

Forrest, T. R. (1978). Group emergence in disasters. In: E. L. Quarantelli (Ed.), *Disasters: Theory and Research* (pp. 105–125). London: Sage.

Fritz, C. E., & Mathewson, J. H. (1957). *Convergence behavior in disasters: A problem in social control*. Committee on Disaster Studies. Disaster Research Group.

Kartez, J. D., & Kelley, W. J. (1988). Research-based disaster planning: Conditions for implementation. In: L. K. Comfort (Ed.), *Managing Disaster: Strategies and Policy Perspectives* (pp. 126–146). Durham, NC: Duke University Press.

National Opinion Research Center (1954). Summary of findings of the Arkansas tornado study. Chicago, IL: University of Chicago/National Opinion Research Center.

Neal, D. M. (1992). Issues of donations following hurricane Andrew. International Sociological Association's Disaster Research Meeting.

Neal, D. M. (1994). The consequences of excessive unrequested donations: The case of hurricane Andrew. *Disaster Management*, *6*, 23–28.

Office of the Inspector General (2002). Update on operation "Vulture Sweep": 23 individuals are charged and 18 arrested for fraudulent World Trade Center claims totaling $760,465. *Eye on OIG*, *2*(8). http://www.ssa.gov/oig/executive_operations/eyeonoig/eyeoig04152002.htm. Last retrieved November 23, 2002.

Quarantelli, E. L. (1982). Human resources and organizational behaviors in community disasters and their relationship to planning. Disaster Research Center Preliminary Paper No. 76. Newark, DE: University of Delaware.

Saad, L. (2001). Americans clearly support a military response to terrorist assault: About half think government could have done more to prevent attacks. The Gallup Organization. Poll Analyses, September 13. http://www.gallup.com/poll/Releases/Pr010913.asp. Last retrieved December 28, 2001.

Scanlon, J. (1992). The man who helped Sammy Prince Write: Dwight Johnstone and the Halifax explosion. *International Journal of Mass Emergencies and Disasters*, *10*(1), 189–206.

Stallings, R., & Quarantelli, E. L. (1985). Emergent citizen groups and emergency management. *Public Administration Review*, *45*, 93–100.

Tierney, K. J., Lindell, M. K., & Perry, R. W. (2001). *Facing the unexpected: Disaster preparedness and response in the United States*. Washington, DC: Joseph Henry Press/National Academy Press.

CONCEPTUALIZING RESPONSES TO EXTREME EVENTS: THE PROBLEM OF PANIC AND FAILING GRACEFULLY

Lee Clarke

ABSTRACT

Disaster and calamity are extreme events that can be used to glean general lessons about how society works. I use the problem of panic to develop several ideas. Panic, we know from years of disaster research, is quite rare at least in the United States. I consider the implication of this for theories of social behavior and human nature. I also suggest the idea of "failing gracefully" as a systems-level notion that highlights the social context of behavior rather than individual panic. I reconsider findings concerning "altruistic" and "corrosive" communities. I critically evaluate the idea of "moral panic," and end with a consideration of the rhetoric functions of "panic."

INTRODUCTION

In 1920, Samuel Henry Prince, who had worked under the tutelage of Frank Giddings, published his doctoral dissertation, *Catastrophe and Social Change* (1920). Empirically, it was a study of the Halifax disaster, which ensued after two ships carrying munitions and volatile chemicals collided. As with all "worst cases," people described the event with superlatives. It created the "largest" explosion ever. It involved the simultaneous occurrence of "impossible" failures. It was

Terrorism and Disaster: New Threats, New Ideas
Research in Social Problems and Public Policy, Volume 11, 123–141

ISSN: 0196-1152/doi:10.1016/S0196-1152(03)11008-3

the "most horrible" thing that Halifax residents had ever seen. *Catastrophe and Social Change* was a landmark work for bringing systematic scholarly analysis to bear on disaster. But Prince's conceptual intervention was more significant than his empirical one. That intervention was less his specific contention that catastrophe brings "a sudden chance for permanent social change" (Prince, 1920) than his general approach of gleaning sociological insight from extreme events.[1]

Twenty years later Pitirim Sorokin (1942) made that approach broader and deeper by trying to use the great range of catastrophes that befall people to understand forms of psychic orientation and the shapes of social change. W. I. Thomas (1909) had ventured a similar effort, and others who followed them would claim that the study of disaster and calamity provided opportunities to discover interesting things about society. But as sociology splintered and specialized, the idea that catastrophe could inspire broad insights seems to have been lost, though some professional disaster researchers always knew it was there. Still, it is generally true that disasters and perceived disasters have come to be seen as special and exotic. If they *are* special then the subfield of "disaster research" is narrow indeed because disasters, however important, are rare. And social science has an understandable bias in favor of focusing on central tendencies rather than anomalies.

But there are at least two reasons to reject that view. First, while catastrophes may be relatively rare, system failures are not. It is normal that governments fail to govern, justice systems become corrupt, educational systems certify rather than educate, and so on. More specifically, there are hundreds of failures in the U.S. air traffic control system every year, scores of "off-standard" conditions in nuclear power plants,[2] and scores of train wrecks with toxic substances aboard. Billions of dollars are spent annually in government efforts to relieve suffering from natural disasters, and tens of thousands die every year in the automobile transport system. And that's in the rich, relatively safe, country of the United States. Social, organizational, and technological systems are far more dangerous and prone to failure in poorer countries. Failure is pervasive and prosaic, not episodic and exotic (Clarke & Perrow, 1996; Zucker & Meyer, 1989).

The second reason not to define failures and disasters, and responses to them, as warranting only narrow attention is the potential theoretic payoffs. This happens with other areas of research. We learn general lessons about flows of information and influence from research on corporate interlocking directorates (Mizruchi, 1996). We learn about how social location shapes worldview from work on abortion activists, stratification, and cults. Whole new conceptions of social structure arise from studies of the creation of professions, the status changes of bureaucrats, and even from the very absence of relationships.

My main goal in this paper is to use instances of catastrophe and calamity as occasions to examine some of the ways that social contexts shape how people respond to risk and uncertainty. How people respond to risk and uncertainty is a central question in social theory; sociologists, especially of risk and disaster, are well-positioned to address it. Other scholars have found it particularly interesting, which is to say theoretically relevant, to investigate how people make choices when faced with risky or uncertain situations. In cognitive psychology, a long line of research and theorizing focuses on the errors in logic and thinking that people commit. One robust finding is that if people are presented with scenarios of probabilistically equivalent choices, but one choice suggests a likelihood to *lose* while the other suggests a likelihood to *win*, they will choose the latter. The people in these experiments seem to interpret the negatively framed choice as worse than the positively framed one, though it is illogical. In economics, there is some disagreement over definitions of "risk" and "uncertainty," but not over their conceptual centrality to economic theory.

Sociologists certainly have not ignored the issue of responses to risk and uncertainty. To mention but a few examples, Stinchcombe (1990) argued that organizations expand in the direction of uncertainties that might impede effectiveness. Short and Clarke (1993) brought together many different kinds of work that hinge on how professionals and organizations deal with risks and uncertainties. Jasanoff (1994) uses "risk" as a lever into issues of technology, inequality, interpretation, and even morality plays.

My intent is not to supplant but to complement existing scholarship by focusing on how people respond to extreme events. I do so by concentrating particular attention on the idea of "panic," and on situations that we might expect to produce panic. My treatment of the idea of panic is part of a broader effort to understand how people build what they consider to be worst cases, and how they respond to situations of social breakdown.

There are more modern examples of using doom and disaster as opportunities to glean general lessons about society. Klinenberg (2002), for instance, uses the 1995 heat wave in Chicago, which killed over 700 people, as an opportunity to analyze and denounce America's inequitable, and unfair, social stratification system. Situations of social breakdown telescope our attention on interesting parts of mind and (re)action. The Los Angeles riots (or upheavals or uprisings) are thought to tell us something about African Americans' connections to the institutions of that troubled city, and their sense of despair at being excluded from them. Revolutions are said to reveal important aspects of the nature of the state or the development of identity. Stories about the breakup of the traditional family are believed to hold messages about changing gender relations and shifting definitions of morality and deviance. Durkheim peered into the abyss of suicide in search of an answer to the broader question, "What are the bonds which unite men one with another?"

When people confront severe disruptions of their environments – when catastrophe strikes, or when people perceive significant degradation of life and environment – their cognitive machinery and their social networks are challenged to find ways to re-establish mental order and collective stability. How people think of that which threatens them affords a glimpse of how they conceive of boundaries and essences. How people respond when their lives and communities are torn asunder reveals the horizons of imagination and the practicalities of effort – whether they can conceive of ways to rebuild and whether those ways can be brought to fruition. All of these issues are basically about how people respond to situations of threat or potential threat.

As a starting generality we can note that the variation in response from individuals is broad, ranging from apathy through bewilderment to rage and zealous protectionism. As well, some people seem to go through stages of deep emotional upset, while others do not (Raphael (1986) reviews all these responses).

Yet while individual-level responses may be broad, the range of responses from collectivities is more narrow. Perhaps it is worth nothing that before such responses are possible, situations must first be defined as threatening. In some cases this involves the realization that a previously safe context now poses grave danger. This is the situation of people surprised to discover that their house is on fire. In other cases the definitional task takes longer and is fraught with more ambiguity, as when a community's groundwater has been contaminated with toxic chemicals. In still other instances there may be little or no physical threat at all, as when people think that suburban malls are crawling with would-be child abductors. Whatever the instance of threat or perceived threat, actors must first define the situation as threatening, and once that happens they are fundamentally facing a risky situation. Perhaps that situation is, even, a potentially worst case.

Here, I examine *panic* as a concept, differentiating its subjective and objective aspects. That distinction suggests the utility of seeing three dimensions of collective panic, each involving a relational failure: network failures, moral failures, and cognitive failures. Next I revisit some extant evidence on responses to gravely threatening situations, including individual and collective responses to catastrophe.

PANIC REACTIONS TO EXTREME EVENTS

There is a common-sense answer to the question, "How do people respond when they confront deeply threatening situations?" Reporters and editors in the media, along with political and corporate officials, are quick to look for public irrationality, especially in the form of panic. There is some support from academia, too, for the expectation that irrational panic will appear. Sociology's strongest case for the existence and importance of public irrationality concerns "moral panics." A widely cited book on the subject, for instance, finds that anti-nuclear power sentiment in the United States is "exaggerated concern," "groundless," and indicative of "panic" (Goode & Ben-Yehuda, 1994).

But what is panic? Surely one element of panic is an overwhelming sense of terror, a feeling of overpowering fear. Many people have that reaction when their airplane hits turbulence. But panic is more than that. If the airplane has to make an emergency landing, and catches fire as it does so, the doors will pop open and perhaps people will climb over their seats and their fellow passengers to exit the aircraft. That rush to exit would probably be called a panic, too.

Abstractly, we can define *panic*, with the Oxford English Dictionary, as an "excessive feeling of alarm or fear . . . leading to extravagant or injudicious efforts to secure safety." The essence of panic, in this conception, is that people take actions that are disproportionate to the actual threat. These actions are often aimed at self-preservation, but have the paradoxical consequence of leading to harm of the individual who panics.

More concretely, "panic" can refer to an emotive and cognitive state of subjectivity in which people *feel* out of control; it can also refer to excited behavior as people try to put distance between themselves and the threatening agent. Psychology is a natural home for such an analysis of panic. Goldenson (1984), cited in Johnson (1987a), calls it "a reaction involving terror, confusion, and irrational behavior, precipitated by a threatening situation." There is a psychological syndrome, for instance, called "panic disorder," which millions of people are said to have (see e.g. Dumont, 1996). "During a panic attack," says a therapist who draws on and contributes to the scholarly literature, "the entire being is under siege. The mind is overwhelmed by thoughts and images of disaster . . ." (Dumont, 1996). People who suffer panic attacks develop feelings of vulnerability that so are strong that they will go to extraordinary lengths to avoid those feelings. Aquaphobics, for instance, are flooded with fear at the prospect of venturing near water. A panic attack gives one an overwhelming sense of doom. For psychology, panic is chiefly an intrapsychic state, existing inside people's heads or in their feelings. The overwhelming sense of doom that panic disorders bring is, we might say, out of proportion to the real danger posed by the thing that's feared. The value of psychology's treatment of panic is chiefly in illuminating the phenomenon's subjective aspects.

Subjectively, panic is a sense of lack of control over self and circumstance, producing a feeling of futurelessness. At an individual level one possible result of such a sensibility is uncontrolled flight. At the collective level the result would likely be seen as a stampede, like animals trying to escape a forest fire. This was said to describe the two tragedies in February 2003, at nightclubs that caught fire in Chicago and Rhode Island. At all levels, from the individual through the collective, the common thread might be called, simply, thoughtless action.

Collectively, this thoughtless action would entail an implosion of social obligation, wherein moral reciprocity and social responsibility are forsaken, giving rise to a hyper-self interested individualist that values only its own survival or betterment. Vivid examples of uncontrolled flight as thoughtless action are in most disaster movies. People push each other aside to save themselves. They take up arms against their neighbors. Well socialized Jekyls turn into raving Hydes.

Such a conception of panic entails a view of human nature not unlike that of neo-classical economics. The true, essential self is fundamentally anti-social because it values self-advantage above all else. Rules, norms, and obligations may be real enough, but they are luxuries of the civilized life. If the proper conditions prevail these luxuries fall away, revealing the inner core of humanity as it really is, ego-centered and selfish. That is, as Williamson (1975) said in laying the foundations for modern transaction cost economics, "human nature as we know it."

But the core of humanity could be quite selfish and there would still be value in distinguishing between the subjective aspect of panic and the objective aspect of it. Disaster research pioneer Quarantelli (1954) wrote the first systematic, sociological treatment of the idea of panic, and basically defined it as a loss of self control. Yet Quarantelli's emphasis was on how people actually behave, and especially on the collective consequences of individual action, rather than the *feeling* of panic.[3] Of course, it is reasonable to assume that someone who has lost control of their self also *feels* out of control. But the distinction is valuable because the phenomena can vary independently.

Where collective panic happens it involves the breakdown of social order, so that people's understandings of how they are tied to others are upended. Collective panic has several components. There is a *moral* failure so that people pursue their self interest regardless of rules of duty and obligation to others. There is a *network* failure, so that the resources that people can normally draw on in times of crisis are no longer there. There is a *cognitive* failure, in which someone's understanding of how they are connected to others is cast aside. The phenomenon of collective panic, then, would be an indication that social organization had failed in substantial ways.

Interestingly, panic, as manifested in uncontrolled flight under conditions of imminent threat of danger, is empirically unusual. Feelings of panic seem to be common but simultaneous moral, network, and cognitive breakdowns are not. Even in the horrors chronicled by the U.S. Strategic Bombing Survey, cities burn, bodies explode, houses fall down and still people do not panic (see Hersey [1946] 1986; Janis, 1951; USGPO). The Survey's graphic stories of firebombs and especially nuclear attack in Japan demonstrate unparalleled horror and excruciating anguish, but little panic. Moreover, even when panic *does* happen, it is almost never a cause for the majority of deaths in a system failure.

There is more contemporary evidence. Norris Johnson has systematically and rigorously investigated two instances in which we would have expected people to panic.[4] The first happened during Memorial Day weekend in 1977, when 165 people perished trying to escape a fire at the Beverly Hills Supper Club, in Southgate Kentucky. The second happened in December of 1979, in Cincinnati, where 11 people were killed at a Who rock concert. The Supper Club case is not far from the familiar ethics question, "is it immoral to yell 'fire' in a crowded theater?" These are instances in which people have become trapped and can suffer grievously if they do not escape. What more appropriate place, then, to witness a breakdown in the bonds that unite people, where heightened emotion lets loose self interest in the service of self preservation?

There were perhaps 2500 people at the Beverly Hills Supper Club, roughly 1200 of whom were in the Cabaret Room, which had 3 exits. Two of those exits were to the side of the room, and led outside, and one was in the front of the room, and led to another part of the club. All of the deaths were within 30 feet of either of the two side exits (Johnston & Johnson, 1988). When the Club's personnel, having discovered fire in the building, started urging people to leave, a handful of people went to the front entrance while the others started calmly filing out of the other exits. But the people who tried to get out of the front entrance soon ran into smoke and fire, so they returned to the Cabaret Room.

Johnson's accounts show clearly that there was a considerable amount of internal, subjective panic but relatively less external, collective panic. People were initially calm as they lined up at the two exits. But after smoke and fire started pouring into the Cabaret Room, some began screaming and others began pushing. When they saw fire enter the room, people started jumping over tables and chairs to get out, as we might expect.

But notice what they did not do. They did not pick up those chairs and use them to strike people queued up before them. They did not grab the hair of those in front of them and shove them aside in a desperate rush to get out. They did not overpower those more helpless than themselves in the interests of self preservation. There were not simultaneous network, moral, and cognitive failures.

FAILING GRACEFULLY, PANIC, AND ALTRUISTIC COMMUNITY

In Kentucky it seems that a small proportion of all the exhibited behaviors were recognizable as panic. As important, panic was probably not the cause of any of the deaths. Indeed, had people developed a sense of urgency sooner more people would have gotten out of the building, and fewer would have died. Rather than attributing the disaster to panic, it is more accurate to say that so many died because the socio-technical system was not designed to fail gracefully.

In Cincinnati, eleven people were killed in a crush that was widely regarded as a panic. The image was that of drunken or drug crazed teenagers who were so desperate to get into a rock concert that they would even trample each other to death. The reality, though, was far different. There were perhaps 8,000 people waiting for the concert. "When the doors opened," writes Johnson, "and the surge to enter began, approximately 25 people fell, some lying on the concrete for as long as 30 minutes" Johnson (1987b). But direct evidence from on-scene witnesses suggests little panic, in the usual sense of that term. In fact, people tried to form a "cordon" around those who had fallen but the push of the crowd behind them was too strong. For the people at the head of the line, the doors in front of them were not an entrance at all, but an *exit* – they were trying to get away from the crowd not inside to see The Who. And as with the Beverly Hills Supper Club, the predominate reports from participants and eye-witnesses are not those of people climbing over each other's backs to escape danger, but of people trying to help others to escape (Johnson, 1987b).

Like the Beverly Hills Club, Cincinnati's Riverfront Coliseum was a system designed with low tolerance for human deviation. It was not designed to fail gracefully. Users of the system would be safe as long as they arrived in anticipated numbers and behaved as designers expected them to. But it was an unforgiving system, little different than a 747 designed to hold 400 passengers but crammed with 800 crashing on takeoff.

The concept of failing gracefully draws our attention to the lager context in which behavior happens. This is not to deny that panic happens; of course panic happens. But systems can be constructed so that they fail so quickly, or in such complex ways, that we should look at attributes of the system for an adequate explanation of failure rather than look to the attributes of the individuals within it.

A burning supper club or an overloaded entrance to a rock concert are situations in which the physical ecology and the social context have degraded precipitously. As noted, they were systems that were not designed to fail gracefully. In software engineering a system that fails gracefully is one in which discrete failures are prevented, by design, from cascading or interacting so that the entire program does not crash. In the present context, the arrangements of social relationships and cultural artifacts (walls, machines, exits, etc.) no longer operate in accordance with the goals and functions for which they were designed. Under such conditions, collective panic should be normal. In fact we would hardly be surprised to discover extremely self-interested behavior, or "unregulated competition" as Johnson calls it, in the service of self preservation. Yet that is not usually the case.

The non-finding of panic is robust. Perhaps the largest available concentration of ideas and research on system failures comes from scholars of natural disasters. After five decades of research on scores of disasters – floods, earthquakes, tornadoes, and all the basic ways that the earth and its environments upset themselves – one of the strongest patterns is that people do not panic. When faults break and the ground shakes, sometimes dwellings crumble and office buildings topple. Such disruptions can crush people (especially poor people, who tend to live in places more likely to crumble) and start raging fires. Yet people hardly ever run screaming through the streets in a wild attempt to escape the terror, even though they are undoubtedly terrorized. Earthquakes and tornadoes come to wreak havoc on neighborhoods or even entire communities. Yet people do not usually turn against their neighbors, or suddenly forget the ties and moral commitments that comprise the threads and fabric of their lives, giving triumph to their natural urges for self interest, self preservation, and self promotion. Instead the more consistent pattern from natural disasters is that people bind together in the aftermath, to restore their physical environments and their cultures to more recognizable shapes.

Rather than panic and paralysis, it is generally true that even when people confront what they consider the worst case, they organize themselves to provide succor and salvation to their friends and even to complete strangers. This is called a "therapeutic community" or "altruistic community" (Barton, 1969). The altruistic community is one in which we see, as Russell Dynes frames it, "the development of an emergency consensus, the development of altruistic norms and behavior, the expansion of the citizenship role, the minimization of community conflict, and the generation of hostility toward outsiders" (Dynes, 1970).

Dynes (1970, Ch. 4; see also Fritz, 1961) gives a systematic explication of how and why an altruistic community develops in response to disaster. The community is normally held together, mechanically and organically, by people with common visions and values, and by a reasonably integrated division of labor. The extreme event temporarily disrupts the division of labor but not the common values. We might say, using my analysis from above, that while there has been a network failure, there have not also been simultaneous moral and cognitive failures. One mechanism that serves as a prophylactic against community disruption is temporary "role expansion," in which people assume a wider and more varied degree of responsibility for others. So the teacher becomes a nurse, the doctor becomes a rescue worker, the mayor becomes a counselor. This structurally induced, broadened sense of responsibility and caring leads people to work together to re-establish a normal and stable social structure. A tornado can level the houses in a neighborhood, but the "neighborhood" remains in the minds and orientations of those who used to live in the dwellings. They are drawn together to work side by side, recovering from an assaulting agent who attacked them all, all the while demonstrating to each other the "natural" character and strength that dwells within but is not normally on display.

CORROSIVE COMMUNITY

Altruism and cohesion, not self-interest and panic, are the central tendencies in response to extreme events. But they are not the exclusive responses. Accumulating evidence suggests there are conditions under which disruptions and disasters do not foster altruistic communities but conflicted ones in which people do not bind together in cooperative reconstruction but split in disagreement and acrimony.

The idea originated in Kai Erikson's analysis of the Buffalo Creek flood (Erikson, 1976). One hundred thirty two million gallons of water were dammed up behind a mining company's slag heap. When the dam failed it became a wall of mud that barreled through a valley and destroyed a community. As noted, previous research on disasters would lead us to expect the formation of an altruistic community, but Erikson found "nothing of the sort" in Buffalo Creek (Erikson, 1976). Instead people developed a more permanent sense of loss, and were often set against one another rather than bound together in a sense of common struggle and recovery. Rather than healing and solidarity he found wounds that would not heal, and conflict rather than cooperation. Erikson generalized his observations to name "a new species of trouble" (Erikson, 1995), where he accorded especial attention to toxic threats. Toxic disasters are a special class that are distinguished by how people respond to them. They have no "frame," he says, which means there is no clear beginning and end to the assault. For example, while we know that being exposed to nuclear radiation or toxic chemicals can be dangerous, unless the exposure is massive or extreme, it is often difficult to determine if actual damage has been done. Collectively, such responses can create a corrosive community.[5]

When communities become corrosive, members' sense of trust in each other and in social order falters. Instead of turning toward one another in solidarity, they turn away from one another in hostility. Relations become caustic, social connections don't reform. Kroll-Smith and Couch (1990) had similar findings in their study of the effects of a long-burning mine fire in Pennsylvania. Freudenburg and Jones (1991), who coined the term corrosive community, also emphasized that technological disasters tend to create more conflict than do natural disasters. People argue (especially through lawyers) about who to blame, or even whether they've been damaged. Neither of those conditions generally hold after a natural disaster. For example, consider work on the effects of the *Exxon Valdez* oil spill by Duane Gill and Steven Picou (1998); see the citations in that paper for references to more of Gill and Picou's important work on the spill). Their meticulous research has shown that people's sense of contamination has worked its way into local communities, so much so that is has been possible to quantitatively document "chronic stress." One result of that stress is that communities, conceived as systems of social relationships, have failed to regain vitality and cohesion, even ten years after the spill.

Other studies corroborate these findings and insights. Edelstein (1988) investigated the effects of contamination on psychological uncertainties and their effects on the formation of community. Shrader-Frechette (1993) showed how experts' fiat regarding the risks of disposing high level nuclear wastes contributed to ambiguous sensibilities for citizens in the American southwest. Egginton (1980) detailed the terribly corrosive effect that an accidental contamination of animal feed had on many rural communities in Michigan. Those effects are also evident in Whiteside's (1979) telling of a dioxin accident in Italy, Clarke's analysis of a chemical accident in New York (1989), Levine's (1982) analysis of Love Canal, and Brown and Mikkelsen's (1990) important work on the contamination in Woburn.

The theoretical and empirical work on altruistic and corrosive communities provides evidence from a macro level on how collectivities respond to threats and perceived threats. Johnson's careful work not only adds more cases of non-panic to existing research but provides the basis for distinguishing between network failures, moral failures, and cognitive failures in cases of system breakdown. Taken together, all of the work reviewed above raises broader questions that revolve around issues of risk and uncertainty. Next I turn to some of those issues.

WHAT DOES PANIC DO?

I began this essay with the entreaty that failure and breakdown not be seen as special. Sudden and dramatic changes in the circumstances of people's lives are opportunities to investigate some (not all) facets of how people think about their connections to each other and how they create something they recognize as communities. The value of seeing failure as normal becomes clear when we answer the obvious question concerning the evidence on panic: Why do people so rarely panic when their physical existence is under grievous threat?[6]

There are two aspects to the answer. First, general rules and norms for how people are supposed to behave stave off panic. Panic rules are not qualitatively different from those that govern people's lives when they *do not* face imminent danger.

When people entered the Beverly Hills supper club they did not check society at the door, along with their coats. The same regulative norms and cultural precepts that people enact and carry around with them every day do not suddenly dissipate, revealing the true, self interested monad when smoke billows into a room. One set of rules that forestalls panic results from the fact that people go to social functions with members of their primary networks. When danger arises the norm, even the instinct, is for people to help their loved ones before they help themselves. At the Supper Club fire as well as The Who concert, first hand reports from concert goers and police showed that people first helped their friends and family, because those were the people emotionally and physically closest to them (Johnson, 1988). When fires start, or viewing stands start to collapse, or the crowd becomes too compact it is someone's parent, child, or lover who is most clearly and immediately at risk. Primary network bonds then are key in preventing the feeling of subjective panic from becoming the behavior of collective panic.[7] People's networks did not fail, and so solidarity remained. Their sense of where they fit in those networks remained, because there was no failure of their cognitive orientations. An overriding self-interest did not emerge, because there was no overwhelming moral failure. There are rules for panic, and people generally follow them. They are not special rules. They are the same rules that govern behavior when people visit the community swimming pool; they are the same operative rules when the theater is *not* on fire.

Second, people who go to such events are probably more alike than different. And at least judging from available disaster research people seem most selfless – though clearly not exclusively so – when they are with others whom they perceive to be like them, most especially in the sense of facing a common future. So that when someone says that "total strangers probably saved my life" (in Johnson, 1987b, p. 367), they are probably wrong. Were the victims and life-savers to find themselves at a backyard cookout, they would probably have quite a lot in common. They would likely be from the same social class, of the same race, and of the same cultural sensibilities. As with the Supper Club fire, Johnson notes a great deal of cooperative behavior at The Who incident, even among strangers. He says that "behavior within the crowd continued to reflect both a social structure comprising the small groups in which people arrived and to conform to sex-role norms" (Johnson, 1987b, p. 370). Social similarity helps forestall the vaulting of self interest seeking with guile.

Analyzing the issue of panic raises the broader sociological problem of proportionality. Epistemologically, panic entails a judgment about the discrepancy between objective threat and internal assessment of that threat. For example, one could say, and many do, that people in contaminated, corrosive communities are in a panic. They are forming opinions and feelings that are out of phase with objective indications of risk and damage. Residents' fears of Love Canal and of Three Mile Island are two famous cases of massive and often corrosive response, but with little or no objective health damage (on Love Canal see Mazur, 1998).

Scholars who use the metaphor of "moral panic" have given serious thought to the issue of proportionality. Generally, the idea of moral panic is used to explain why people fear phenomena – ethnic groups they are not members of, technologies, crime, etc. – in ways that are "disproportionate" to the actual harm they cause (Cohen, 1972).[8] Goode and Ben-Yehuda have written the most systematic treatment of the idea of moral panic. Like others, they talk of "erroneous beliefs," "irrationality," and of course "panic." They note that the notion of panic depends on "disproportionality." This is so whether the referent is panic in a disaster or a moral panic in collective behavior. As they say, "the term moral panic conveys the implication that public concern is in excess of what is appropriate if concern were directly proportional to objective harm" (Goode & Ben-Yehuda, 1994, p. 36; Steve Rayner & Cantor, 1987). Similarly, in *Moral Panic*, Jenkins (1998) examines the historical variation in the evaluation of behavior we now call "sex crimes." Borrowing the moral panic idea from Cohen (1972), Jenkins (1998, pp. 6–7) analyzes the "extravagant claims-making by professionals," "irrational public fear," and "panic responses." We learn that sometimes public concern over an issue is "logical and comprehensible," as when fears over sexually transmitted diseases started to rise in the 1980s after the discovery of incurable diseases. But in other cases, "the perceived significance of a given problem grows or diminishes without any change in the real threat-potential of the condition itself" (1998, p. 3).

Yet the idea of proportional response (known as "accurate risk perception" in risk-studies) is more intractable than has been appreciated. Whether some set of behaviors is or is not a *panic* depends on establishing that the mismatch between perception and reality is true, and doing *that* seems to require a very low or non-existent level of uncertainty about that which is perceived as threatening. This is so because the threatening phenomenon must be known in a valid and reliable way. This should be a fairly straightforward procedure in some instances. For example, it would not require much effort to know the actual risk from witches casting demonic spells on innocent children. However, establishing lack of correspondence between perception and probability is much less straightforward in technical disputes. The problem of uncertainty in knowledge of threats must be attended to before a judgment of panic is warranted. For example, we must know that the probability of a nuclear power plant melting down is, say, one in a million reactor years. Without such evidence, the judgment of disproportionality would likely rest on speculative foundations.

Moreover, a central metric, especially in technical conflicts, by which to judge that people have responded disproportionately to some perceived threat is that their perceptions differ significantly from those of experts who exhibit considerable agreement on the probability of threat (Goode & Ben-Yehuda, 1994, pp. 37–38). But that may be a poor metric to use. First, it fails to recognize errors and biases of experts, even in their own domains of knowledge. Studies of corporate managers (March & Shapira, 1987; Mintzberg, 1973), government regulators, and technical experts (Freudenburg, 1988; Lichtenstein & Fischhoff, 1977; Mazur, 1973) show they are prone to extra-rational biases such as organizational pressures and professional ideologies (Dietz & Rycroft, 1987; Vaughan, 1996). A series of case studies demonstrate how experts and decision makers can become intransigently committed to beliefs about what is and is not risky, while non-experts seek new information in uncertain situations (Brown & Mikkelsen, 1990; Edelstein, 1988; Levine, 1982).

Second, relying on expert agreement privileges expert epistemologies that rely on probability theory and cost-benefit logic, while neglecting less officially sanctioned ways of knowing. From the view of probability theory, that which is most important to worry about is that which is most likely. Since worst cases or extreme events, especially catastrophic ones, are always highly unlikely, people who worry about them would by the lights of moral panic theory be irrational. Yet just because catastrophic failures, or even less severe degradation of life and environment, are statistically rare does not make them trivial. It is common, and true, to say that airliner travel is the safest part of any trip. But for any given individual, standing at the door of a plane at the beginning of the flight, the probability of a successful flight (or landing) is close to zero or one. From *that* point of view, it is far from irrational to worry about the safety of airline travel.

Finally, benefit/cost reasoning adequately registers damages that are obvious such as body counts and morbidity rates. But other metrics may be just as useful. Perrow (1984) suggested that people may be "socially rational," which is a more complex conception of the risk calculus. Rather than panic or disproportionate (aka illegitimate) fear, people may worry less about obvious damage and more about catastrophic potential because the latter wreak social, cultural damage. Work on natural hazards and technological risks and uncertainties affirms the importance of social disruptions – to community structure and ethos, hopes and expectations for family futures, institutional legitimacy, collective responsibility – that are hard for markets to price, hence hard to include in a benefit/cost analysis. From this view, dismissing non-expert fears as panics neglects the larger social contexts in which people's lives are embedded (Freudenburg, 1988).

CONCLUSION

Real, collective panic – involving simultaneous network, cognitive, and moral failures – is rare. Even in places where we most expect it – the proverbial and metaphorical burning theater – true panic seems to happen only when fire is climbing up people's backs and smoke has begun to choke the life out of them. Even when panic *does* happen it seems to be accompanied by heroism and altruism. Furthermore, genuine panic is rarely a major cause of deaths. Indeed, people seem to keep their behavioral wits about them even when their emotional impulses are urging self-preservation and selfishness. Repair thy neighbor is more the rule when calamity strikes than revere thy self. Thus, in one sense the message here is the usual one from sociology – things are more complicated than they look. Or, more technically, we could say that contexts in which harm is extreme and imminent elicit from people a variety of responses, only one of which is panic, and that one is uncommon. In most situations where panic is alleged to be the cause of death and destruction, the idea that systems are not designed to fail gracefully is a better description.

Panic is both a simple and a complex idea. It is simple because it is a single word that apparently characterizes a wide range of situations. It is complex because it serves multiple purposes, purposes that are sometimes contradictory. Panic is more interesting in its role as complex idea, but more important, in a sense, in its role as a simple idea. Panic's multiple purposes include its role as a sociological concept, its role as a political diagnosis, and its role as a rhetorical device in power struggles.

Panic is conceptually sharp as a political tool. The subtext in most announcements that some group has panicked is that the alleged panikers (always lay people) do not value what elites value. As a political device "panic" works well as part of a strategy of what Freudenburg calls "diversionary reframing" (Freudenburg et al., 1988). Strategies of diversionary reframing are often used by elites in technical debates as ways to change the nature of a conversation in which the official conception of reality is being challenged. If lay-people begin arguing with officials about the dangers and unfairness of toxic contamination, for example, a frequent response is to question the credentials of the challengers. Doing so reframes the debate so that the issue is no longer about contamination but about who knows best what the "real" danger is. Similarly, experts in the nuclear industry have always been fond of pointing out that far more people die every year in prosaic accidents (e.g. involving lawn mowers or bathrooms) than from atomic power plant accidents. Of course people (American people at least) are more afraid of nuclear plants than lawnmowers. In this clear discrepancy in opinion, pro-nuclear experts have morbidity and morality rates on their side. The logical conclusion is that the non-experts are panicky, their fears out of proportion to what experts know to be the real danger. People who strongly disagree with the message of their opponents tend to level at them a charge of unreasonableness or even irrationality. To say of those protesting toxic threats that they are in a panic is to put them at a rhetorical disadvantage.

As a simple idea, panic resonates as a vocabulary of motive, more so than as a driver of social action. "Motives are of no value," thought C. Wright Mills, "apart from the delimited societal situations for which they are the appropriate vocabularies" (1940, p. 11). Motive-mongering as a form of explanation, in other words, is unenlightening. From that vantage point, "panic" functions as a rhetorical device, of greater or lesser utility in convincing various audiences that a situation of catastrophic failure has been "explained." Perhaps the key to reconciling apparently competing ideas about panic and non-panic, rationality and irrationality, or corrosive and altruistic communities is to back away from the behaviors themselves and conceptualize the institutional contexts that provide safe haven for such attributions.

NOTES

1. For comments thanks to David Guston, James Jasper, Charles Perrow, Patricia Roos. I used some of this material in Clarke (2002).
2. See http://www.nrc.gov/OPA/primer.htm
3. Quarantelli (2001) has recently come to the conclusion that the idea of panic has no analytic utility at all.
4. I'm drawing from a series of articles, including Johnson (1987a, b, 1988) and Feinberg and Johnson (1995), and Johnson et al. (1994).
5. Obviously, Buffalo Creek was not a toxic threat. What that case has in common with toxic ones is that they are perceived to be caused by human recreancy. My concern here is with the range of responses to extreme events rather than a classificatory debate.
6. Quarantelli points out that even when people do flee in panic they do not do so in a "random or helter-skelter" way (1954, p. 269). Panic flight, he says, is functional. If, responding to calamity, people fell into a wildly uncontrolled panic then we would expect more randomness in the *forms* of their responses. But, as Quarantelli quipped, the person in panic "does not run blindly into a wall; he heads for a door" (1954, p. 272).
7. Johnson also found that *workers* at the Supper Club did not place self-interest over the safety of others. The same can be said of the famous Cocoanut Grove conflagration in Boston in 1942. Nearly 500 people died in that fire and while it is clear that there were pockets of panic, it is equally clear that it was not panic that produced the high body count. It was, rather, another instance of an overloaded system that could not fail gracefully.
8. James Jasper (personal communication) argues that the force of work on moral panics is that it emphasizes how elites manipulate popular sentiment and perception, not that every-person is prone to irrationality. But if there is no public irrationality then where is the "moral panic?"

REFERENCES

Barton, A. H. (1969). *Communities in disaster: A sociological study of collective stress situations.* Garden City, NY: Doubleday.

Brown, P., & Mikkelsen, E. J. (1990). *No safe place: Toxic waste, leukemia, and community action.* Berkeley: University of California Press.

Clarke, L. (1989). *Acceptable risk? Making decisions in a toxic environment.* Berkeley: University of California Press.

Clarke, L. (2002). Panic: Myth or reality. *Contexts, 1*(3), 21–26.

Clarke, L., & Perrow, C. (1996). Prosaic organizational failure. *American Behavioral Scientist, 9*(8), 1040–1056.

Cohen, S. (1972). *Folk devils and moral panics: The creation of the Mods and Rockers.* London: MacGibbon and Kee.

Dietz, T., & Rycroft, R. W. (1987). *The risk professionals.* New York: Russell Sage Foundation.

Dumont, R. (1996). *The sky is falling: Understanding and coping with phobias, panic, and obsessive-compulsive disorders.* New York: W. W. Norton.

Dynes, R. R. (1970). *Organized behavior in disaster.* Lexington: D.C. Heath.

Edelstein, M. A. (1988). *Contaminated communities: The social and psychological impacts of residential toxic exposure.* Boulder: Westview Press.

Egginton, J. (1980). *The poisoning of Michigan.* New York: W. W. Norton.

Erikson, K. (1976). *Everything in its path: Destruction of community in the Buffalo Creek flood.* New York: Simon and Schuster.

Erikson, K. (1995). *A new species of trouble.* NY: Norton.

Feinberg, W. E., & Johnson, N. R. (1995). Firescap: A computer simulation model of reaction to a fire alarm. *Journal of Mathematical Sociology, 20*(2–3), 247–269.

Freudenburg, W. R. (1988). Perceived risk, real risk: Social science and the art of probabilistic risk assessment. *Science, 242*(October), 44–49.

Freudenburg, W. R., Frickel, S. F., & Dwyer, R. E. D. (1988). Diversity and diversion: Higher superstition and the dangers of insularity in science and technology studies. *International Journal of Sociology and Social Policy, 18*(5/6), 6–34.

Freudenburg, W. R., & Jones, T. R. (1991). Attitudes and stress in the presence of technological risk: A test of the supreme court hypothesis. *Social Forces, 69*(4), 1143–1168.

Fritz, C. E. (1961). Disaster. In: R. K. Merton & R. A. Nisbet (Eds), *Contemporary Social Problems.* New York: Harcourt, Brace and World.

Gill, D. A., & Picou, S. (1998). Technological disaster and chronic community stress. *Society and Natural Resources, 11,* 795–815.

Goldenson, R. M. (Ed.) (1984). *Longman dictionary of psychology and psychiatry.* New York: Longman.

Goode, E., & Ben-Yehuda, N. (1994). *Moral panics: The social construction of deviance.* Cambridge: Blackwell.

Hersey, J. ([1946]1986). *Hiroshima.* New York: Bantam Books.

Janis, I. L. (1951). *Air war and emotional stress.* New York: McGraw-Hill.

Jasanoff, S. (1994). *Learning from disaster.* Philadelphia: University of Pennsylvania Press.

Jenkins, P. (1998). *Moral panic: Changing concepts of the child molester in modern America.* New Haven: Yale University Press.

Johnson, N. R. (1987a). Panic and the breakdown of social order, popular myth, social theory, empirical evidence. *Sociological Focus, 20*(3), 171–183.

Johnson, N. R. (1987b). Panic at "The Who concert stampede": An empirical assessment. *Social Problems*, *34*(4), 362–373.

Johnson, N. R. (1988). Fire in a crowded theater: A descriptive investigation of the emergence of panic. *International Journal of Mass Emergencies and Disasters*, *6*(1), 7–26.

Johnson, N. R., Feinberg, W. E., & Johnston, D. M. (1994). Microstructure and panic: The impact of social bonds on individual action in collective flight from the Beverly Hills Super Club fire. In: R. R. D. Dynes & T. Kathleen (Eds), *Disaster, Collective Behavior, and Social Organization* (pp. 168–189). Newark: University of Delaware Press.

Johnston, D. M., & Johnson, N. R. (1988). Role extension in disaster: Employee behavior at the Beverly Hills Supper Club fire. *Sociological Focus*, *22*(1), 39–51.

Klinenberg, E. (2002). *Heat wave: A social autopsy of disaster in Chicago*. Chicago: University of Chicago Press.

Kroll-Smith, J. S., & Couch, R. S. (1990). *The real disaster is above ground: A mine fire and social conflict*. Lexington: University of Kentucky Press.

Levine, A. (1982). *Love Canal: Science, politics, and people*. Lexington: D.C. Heath and Co.

Lichtenstein, S., & Fischhoff, B. (1977). Do those who know more also know more about how much they know? *Organizational Behavior and Human Performance*, *20*, 159–183.

March, J. G., & Shapira, Z. (1987). Managerial perspectives on risk and risk taking. *Management Science*, *33*(11), 1404–1418.

Mazur, A. (1973). Disputes between experts. *Minerva*, *11*(2), 243–262.

Mazur, A. (1998). *A hazardous inquiry*. Cambridge: Harvard University Press.

Mills, C. W. (1940). Situated actions and vocabularies of motive. *American Sociological Review*, *5*(6), 904–913.

Mintzberg, H. (1973). *The nature of managerial work*. New York: Harper and Row.

Mizruchi, M. S. (1996). What do interlocks do? an analysis, critique, and assessment of research on interlocking directorates. *Annual Review of Sociology*, *22*, 271–298.

Prince, S. H. (1920). *Catastrophe and social change, based upon a sociological study of the Halifax disaster*. NY: AMS Press.

Quarantelli, E. L. (1954). The nature and conditions of panic. *American Journal of Sociology*, *60*(3), 267–275.

Quarantelli, E. L. (2001). Sociology of panic. In: P. B. Baltes & N. Smelser (Eds), *International Encyclopedia of the Social and Behavioral Sciences*. Pergamon Press.

Raphael, B. (1986). *When disaster strikes: How individuals and communities cope with catastrophe*. New York: Basic Books.

Rayner, S., & Cantor, R. (1987). How fair is safe enough? the cultural approach to social technology choice. *Risk Analysis*, *7*(1), 3–9.

Short, J. F., & Clarke, L. (1993). Social organization and risk: Some current controversies. *Annual Review of Sociology*, *19*, 375–399.

Shrader-Frechette, K. S. (1993). *Burying uncertainty: Risk and the case against geological disposal of nuclear waste*. Berkeley: University of California Press.

Sorokin, P. (1942). *Man and society in calamity: The effects of war, revolution, famine, pestilence upon human mind, behavior, social organization, and cultural life*. New York: Greenwood Press.

Stinchcombe, A. L. (1990). *Information and organizations*. Berkeley: University of California Press.

Thomas, W. I. (1909). *Source book for social origins*. New York: Gorham.

USGPO. *U.S. strategic bombing survey, the effects of strategic bombing on Japanese morale, moral division, 1947; Strategic bombing survey, Physical Damage Division. Effect of the incendiary bomb attack on Japan*, 1946, pp. 94–97 and SRI, Vol. 1, pp. 93–97. Washington.

Vaughan, D. (1996). *The challenger launch decision*. Chicago: University of Chicago Press.
Whiteside, T. (1979). *The pendulum and the toxic cloud*. New Haven: Yale University Press.
Williamson, O. (1975). *Markets and hierarchies*. New York: Free Press.
Zucker, L. G., & Meyer, M. M. (1989). *Permanently failing organizations*. Newbury Park: Sage Publications.

www.ingramcontent.com/pod-product-compliance
Ingram Content Group UK Ltd.
Pitfield, Milton Keynes, MK11 3LW, UK
UKHW020843290926
15470UKWH00004B/1519

9 780762 310432